FORMULA 1 CURIOSITIES

Volume 1 – 2023 Edition

Text and sketches

by Carlos de Paula

De Paula Publishing Corp.

This work is of a reference nature. The data provided herein has been obtained from multiple sources deemed to be reputable, and where available, it has been cross-checked. I welcome any corrections, which, if substantiated, will be implemented in future editions. Such corrections may be sent to carlosdepaula@mindspring.com. Any text of a subjective nature merely reflects the opinion of the writer. Some statements of fact are quoted from third party works, who assume full responsibility for it. This work is copyrighted, and its content cannot be reproduced in any shape or form, including Facebook or other social media groups, without the express consent of the author.

ISBN: 9798864593769

© 2023, Carlos A. de Paula

Cover Photo: © Bernard Cahier, reproduced with permission

Patrick Depailler in Tyrrell P-34, Long Beach, 1977

Published by De Paula Publishing and Services Corporation
900 Bay Drive, suite 125, Miami Beach, Florida 33141

*I dedicate this book to
the throngs of anonymous
mechanics and support
personnel that made
Formula One possible
since 1950*

Paperback and hard cover available at
motorracingbooks.com

PREFACE

I have been an avid follower of Formula One since 1972 and have written a few books on racing which have been sold in a few countries.

During these years I have accrued quite a lot of information concerning Formula One, dating back to 1950, and decided to publish books on the most curious aspects of this particular niche of the sport, which changed quite a bit since inception.

Having said that, there are two things this book is not: it is not a statistics book and it is not a pictorial book.

To be honest, dry statistics without commentary can be widely found on the internet, most of it correct. Likewise, there are probably hundreds of thousands of F1 photographs on the open internet, Facebook, Pinterest and Instagram, which are published without having to pay photographic rights, due to different treatment afforded to websites under copyright law.

Which is not to say the book does not contain statistics: it does, but they are presented as isolated facts and contextually, which can help you get a picture of the evolution and standing of the sport since 1950.

I have always felt the best way to choose a book subject is to write one I would buy. I have always liked curious facts and love details. I had fun writing this book and hope that you, the reader, will have fun as well.

I believe there is something for everybody here. Old timers will be able to reminisce about the good old times, and perhaps learn some details from latter years that have escaped them, and the scores of newcomers who know very little to nothing about the history of Formula One will certainly be amazed how different the sport was in previous decades. Hopefully, this will whet their appetite, and lead them to further research.

I have divided this book into chapters on specific issues, such as grids, numbers, disappointments, lengthy career spans, etc, with applicable commentary. This is often of a subjective nature, and I know some readers may dislike my views. Unfortunately, it is impossible to please everybody, as I have found out in my previous books. I know some readers may say they have read some of these stories many times over, forgetting that perhaps, the majority of the stories and details have passed far from their radar.

Since I want this to be a series, I did not include all interesting stories and facts, for I want to have enough material for volume II, III, etc., and do not want the book to be very extensive and expensive.

The book is meant to focus on details that may not be very clear to fans, narrated in, I hope, an entertaining fashion. The text also provides a historical perspective, especially interesting for the legion of new fans.

I have tried to give even coverage to all periods. One should bear in mind that the amount of information available from the 50s is not comparable to what is available today, given that seasons comprised of as little as seven races (adding the Indy 500 in the early days).

I have stayed away from private lives of drivers, managers, owners, officials. I am not fond of gossip and titillations, and this book is about racing, not celebrities per se.

To make the book affordable, it is not pictorial. You will find a few photos and sketches here and there. Unfortunately, photographic rights are very expensive these days, and sweetheart deals seem to be available only to members of the automotive press, which is not my case.

I would like to thank Paul-Henri Cahier, Rob Neuzel, Kurt Oblinger, Alejandro de Brito, Russel Whitworth, Lola Heritage, Teddy Pilette, Gerald Swan, Jose Santos, Paul Kooyman, Vicky Chandhok, Jim Llewellyn, Stephen Lawrenson, Martin Fokkens, Jan Borsboom, Pieter Kamp, Paul Maes, Ricardo Cunha, Wagner Gonzalez, Fritz Jordan, Eshan Pieris, Teddy Pilette, Jurgen Barth, Pete Austin, Rob Petersen, Douglas Noordhoorn, Tony Trimmer, Bjorn Lahus, Enrique Soto, Michael Lochmann, Alexander Matveev, Paul Rutten, Jari Debner, Harry Hammaren, Hans Hugenholz, Joey Anastasi, Therry Van Vreden, Ben Osten. Romeu Nardini, Paulo Lava, Rui Amaral Lemos, Roberto Costa, Desmond Troy Hillary, Tony Beukes, Richard Francis, Dave Flick, Wendel Bingham, Ron Mayberry, Todd Bettenhausen, Vic Angle, Scott Fischer, Bruce Smith, Lisa Haight, Jenny Ambrose, Jose Mota Freitas, Jackie Marquart, Darrell Hanestad, , Jeff Walker, Pat de Klerk, Maria do Carmo Moreira, Leon Guisti, Joe Cali, Bjorn Hehmann, Derek Ziman, Mike Geiger, Richard Stringer, Alan Kernick, Terry Porter, Frans Van Der Linden, Francois Pienaar, Mike Wesson, Jovino Benevenuto Coelho, Luis Valverde, Blade de Meillon, Francois Pretorius, Michael Beukes, Sean Woodgate, Vilia Celliers, Kevin Oldfield, Steve Philipps, Jerry Nowlin, Tim Tischendorf, Matt Kendall, Dave Pleus, Mike Jacobs, Carroll Hamilton, Jim Keeker Paul Troy, Hans Tremi, Steve Toit, Pieter Visser for assistance provided in the preparation of my books.

Unattributed photos are deemed to be public domain. Should you own the copyright to any photo, kindly notify me and I will either provide credit or remove it at once.

Carlos de Paula, Miami Beach

Paperback and hard cover available at

motorracingbooks.com

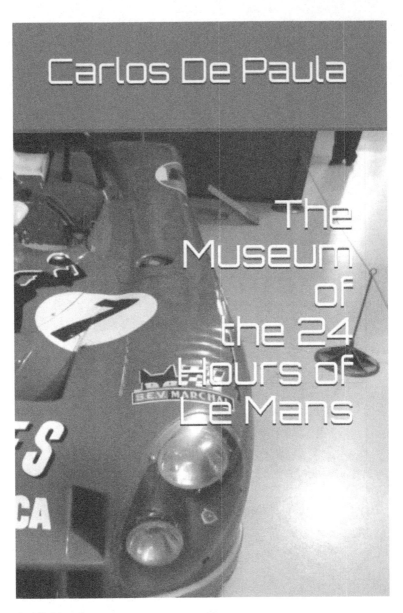

Available at

TABLE OF CONTENTS

Year-by-year highlights, 11

Grids, 84

Car numbers, 88

Sponsors, 90

Engines, 96

Cars with long careers, 99

Nationalities, 104

Driving from the back of the field, 110

Drivers that should have won, 112

Eponymous teams, 118

Nobility, 125

Qualifying surprises..., 130

...and surprising fastest laps, 137

Unique podiums, 143

Ferrari hits and misses, 149

Major disappointments, 154

Team DNA, 163

Dynasties, 172

Future movers and shakers, 193

Insufficient challenges, 197

Statistics, 203

The U.S. Grand Prix saga, 206

Major failures, 211

Great accomplishments, 226

Shared drives, 230

Scary stuff, 235

Lengthy racing career spans, 239

The Grand Prix that did not happen, 256

Major heartaches, 260

Unusual entries, 268

Early bloomers - drivers who won in their debut season, 271

Late bloomers - drivers that took a long time to get their first win, 274

Back and forth, forth and back, 280

World champion sabbaticals, 287

The Indy 500 angle, 295

When Formula One is not really Formula One, 298

A whole bunch of Johns, 304

The Monaco Grand Prix: Glamour, Excitement and Boredom, 306

Where World Champions raced at one time, 313

Bibliography, 318

YEAR-BY-YEAR HIGHLIGHTS

1950

- First FIA World Championship for Drivers. With 6 races in Europe and the Indy 500, the races were contested with Formula 1 cars (4.5-liter normally aspirated, 1.5-liter with supercharger), Formula 2, Indy racers and even occasional sports cars.
- Alfa Romeo won all six Formula 1 races using a pre-war 1.5 Alfa Romeo 158 equipped with a supercharger.
- In addition to Alfa, other constructors making appearances were Talbot, Ferrari, Maserati, in addition to pre-war ERA voiturettes, also supercharged, and a few others.
- Italian Giuseppe Farina was the first champion, and also won the inaugural race in Silverstone, England. He won 3 races, to Fangio's 3.
- First GPs held in England, Monaco, Switzerland, Belgium, France, Italy, plus the Indy 500.
- Nine cars are eliminated in the Monaco Grand Prix, as a large wave crashes over the sea wall. Fangio manages to avoid it, but Farina does not, crashes the car that bounces between sides of the track, obstructing it.
- The Indy 500 is interrupted at 138 laps due to rain.
- First Italian world champion (Farina). First champion from Europe.
- The scoring is: 1st (8 points), 2nd (6), 3rd (4), 4th (3), 5th (2), plus one point awarded for fastest lap. Points were divided for drivers in shared drives.
- First wins by Italian, Argentine and American drivers.

Giuseppe Farina was the first world champion and also won races for Ferrari. (Alejandro de Brito)

- First win by Italian cars. An American car wins the Indy 500.
- First win drivers: Farina, Fangio, Parsons
- First win constructors: Alfa Romeo, Kurtis-Kraft
- Drivers from the following countries drove in this season: Italy, Argentina, Britain, France, Thailand, Belgium, Switzerland, Monaco, Ireland, USA, Germany
- Marques which raced for the first time: Ferrari, Alfa Romeo, Maserati, Alta, ERA, Cooper, Simca-Gordini, Kurtis Kraft, Deidt, Moore, Lesovsky, Nichels, Marchese, Stevens Langley, Ewing, Rae, Olson, Wetteroth, Snowberger, Adams, Watson.

1951

- Fangio's first championship, still driving for Alfa Romeo.
- First Argentine world champion.
- First champion from the Americas.
- Ferrari's first win in the Championship, in the British Grand Prix. The car was driven by Argentine Jose Froilan Gonzalez.
- Longest race ever in the championship: Indy 500, 3 hours 57 minutes and 38 seconds
- First shared win: Fangio/Fagioli in France.
- At the end of the season Alfa quits.
- Fangio is the winningest F1 driver at the end of the season, with a career total of 6 wins.
- First races held in Germany and Spain.
- First win drivers: Gonzalez, Ascari, Fagioli, Wallard
- First win constructors: Ferrari
- Drivers from the following countries drove in this season: Argentina, Italy, Switzerland, Monaco, Britain, Belgium, USA, France, Brazil, Germany, Spain, Thailand
- New nationalities represented: Brazil, Spain

- Marques which raced for the first time: Veritas, HWM, Sherman, Schroeder, Pawl, Bromme, Trevis, BRM, Osca

Alfa Romeo's 158 and 159 won the first two editions of the Championship. This sketch shows the obvious 1930s origin of the car.

1952

- With the absence of Alfa Romeo, the FIA decides the championship will be contested by 2.0 liter Formula 2 cars.
- Ferrari takes full control of the championship, its type 500 winning all races except the Indy 500
- Fangio sits out the season, due to injuries sustained in the Monza Grand Prix. Before that, he won 6 Formula Libre races in South America.
- Taruffi won the season's first race, for team leader Ascari was driving a Ferrari at Indy.
- In the next Grand Prix Ascari begins a straight 9-race win that goes into 1953 and wins his first championship.
- Alberto Ascari becomes the winningest F1 driver at the end of the season, with 8 wins
- First race held in Holland.
- First win drivers: Taruffi, Ruttman
- First win constructors: Kuzma
- Drivers from the following countries drove in this season: Italy, Switzerland, France, Britain, USA, Germany, Belgium, Thailand, Australia, Brazil, East Germany, Netherlands, Argentina, Uruguay
- New nationalities represented: Australia, East Germany, Netherlands, Uruguay

- Marques which raced for the first time: AFM, Kuzma, Frazer-Nash, Connaught, BMW Greifzu, BMW Heck, BME Eigenbau, BMW Special, Gordini, Aston Butterworth, Cisitalia

An army of Ferrari 500 resting before battle (Alejandro de Brito)

1953

- The Indy 500 is held under scorching conditions, and a large number of drivers require relief to finish the race. One of them, Carl Scarborough dies in the field hospital from heat exhaustion. The winner Bill Vukovich drives solo.
- Chet Miller dies in practice for the Indy 500
- Ascari increases his win total to 13.
- First championship win by a British driver, Hawthorn at Spain, driving a Ferrari.
- First race held in a South American country, Argentina
- First win drivers: Hawthorn, Vukovich
- Drivers from the following countries drove in this season: Italy, Argentina, Britain, France, USA,

Switzerland, Belgium, Germany, East Germany, Thailand, Monaco, Brazil
- Marques which raced for the first time: EMW, Turner, Del Roy

1954

- Change to 2.5 liter normally aspirated/750 cc supercharged formula.
- Fangio wins the first race of the season driving a Maserati, then shifts to the Mercedes team.
- Mercedes debuts in the Championship, fielding a streamlined (covered wheel) car at Reims, where Fangio wins.

The covered wheel streamlined Mercedes looked more like a sports car. It won on debut in Reims, 1954

- The Argentine wins his second championship in commanding style. Drivers from Argentina finish 1-2 in the championship, as Gonzalez is the runner-up.
- Fangio ties with Ascari as winningest F1 driver at the time, with 13 wins

- Fangio becomes the first driver to win races for two different constructors in the same season, Maserati and Mercedes.
- Ascari shifts to the Lancia team, where he finds the going tougher.
- For the second year running the Indy 500 is run under very hot and uncomfortable conditions, and many drivers require relief. The winner Vukovich drives solo again.
- First win by a German car.
- First win constructors: Maserati, Mercedes
- Drivers from the following countries drove in this season: Argentina, Italy, France, USA, Thailand, Britain, Germany, Belgium, Spain, Switzerland
- Marques which raced for the first time: Mercedes-Benz, Phillips, Pankratz, Vanwall, Lancia

1955

- Fangio wins the championship again for Mercedes, now with Stirling Moss as his teammate.
- In the wake of the Le Mans accident involving Pierre Levegh, Mercedes quits at the end of the season. Levegh was driving a Mercedes in the race.
- Longest F-1 ever held in the Championship: German Grand Prix, 3 hours 45 m 58 seconds
- First win by a French driver.
- Fangio passes Ascari and becomes the top F1 winner until 1968.
- Mannu Ayulo dies in practice for Indy
- After Ascari's fatal accident in testing, Lancia withdraws. Its equipment is turned over to Ferrari.
- First win drivers: Moss, Trintignant, Sweikert
- Drivers from the following countries drove in this season: Argentina, Italy, France, Germany, Britain, Uruguay, Monaco, Belgium, Brazil, Australia, USA

- Marques which raced for the first time: Epperly, Trevis, Arzani-Volpini

Here is Fangio in the 1955 Mercedes (Alejandro de Brito)

1956

- Bill Vukovich dies at Indy 500. He would be the only 2-time winner of the race while valid for the World Championship, and led 485 laps.
- Fangio and Musso share the winning car in Argentina.
- First win drivers: Musso, Collins, Flaherty
- First win constructors: Lancia-Ferrari, Watson
- Drivers from the following countries drove in this season: Argentina, Britain, France, Uruguay, Brazil, Italy, Monaco, Belgium, Spain, USA, Australia, Switzerland, Sweden, Germany
- New nationalities represented: Sweden
- Marques which raced for the first time: Lancia-Ferrari

1957

- Last shared win, Moss/Brooks (Britain)

- Single Grand Prix held at Pescara. For the first time more than one race is held in the same country (Italy) during a season
- Keith Andrew dies in practice at the Indy 500
- First championship win by a British car (Vanwall, Britain)
- First win drivers: Brooks, Hanks
- First win constructors: Vanwall, Salih
- Drivers from the following countries drove in this season: Argentina, France, USA, Spain, Italy, Britain, Germany, Sweden, Australia, Netherlands, Switzerland
- Marques which raced for the first time: Dunn, Porsche

1958

- First British world champion, Hawthorn. Moss is runner-up for the fourth straight time.
- British constructors dominate the season, but in the end Ferrari wins again. With a British driver...
- First win by a rear-engined car, Moss in a Cooper-Climax in Argentina.
- Moss becomes the second driver to win races for two different constructors in the same season, Cooper and Vanwall.
- Fangio's last season is far from successful. The 5-time champion raced twice in Formula 1, getting 4th places in Argentina and France, driving private Maseratis, and does not qualify for the Indy 500. He actually could not get up to speed and withdrew. Another driver ended up crashing his ride.
- Maria Teresa de Filipis becomes the first woman to drive in the World Championship
- Future Champions Graham Hill and Phil Hill debut in Formula 1.
- Lotus debuts in Formula 1
- Points are no longer awarded for shared drives.

- First constructors cup, winner is Vanwall.
- Stewart Lewis-Evans dies in the Moroccan Grand Prix.
- Luigi Musso dies in the French Grand Prix
- Peter Collins dies in the German Grand Prix
- Pat O'Connor dies in the Indy 500. Jerry Unser and Bob Cortner died in practice.
- Hawthorn retires at the end of the season
- First race in the African continent, in Morocco.
- First win drivers: Bryan
- First win constructors: Cooper
- Drivers from the following countries drove in this season: Britain, Italy, Argentina, France, USA, Spain, Australia, Germany, Sweden, Monaco, Netherlands, Belgium, New Zealand, Morocco
- New nationalities represented: New Zealand
- Marques which raced for the first time: Lotus

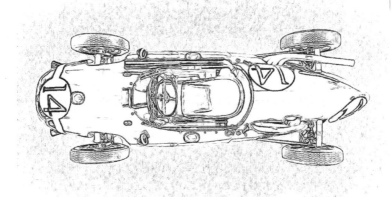

Cooper revolutionized Formula 1 in 1958

1959

- First Australian world champion, Jack Brabham

- First win by an Australian driver.
- First win by a Swedish driver.
- First win by a New Zealand driver.
- First Portuguese Grand Prix.
- 1958 champion Hawthorn dies from a road accident in January.
- First non-Indy 500 US Grand Prix held at Sebring, Florida.
- First time a 2-heat Grand Prix is held. The venue is Avus, Germany and the idea is not repeated.
- First win drivers: J. Brabham, Ward, Bonnier, McLaren
- First win constructors: BRM
- Drivers from the following countries drove in this season: Britain, Australia, France, USA, New Zealand, Sweden, Germany, Italy, Belgium, Netherlands, Brazil, Uruguay, Portugal, Monaco, Argentina
- New nationalities represented: Portugal
- Marques which raced for the first time: Elder, Sutton, Christensen, Moore, Aston Martin, Fry, Tec Mec

1960

- First GP win by an American driver. Until then, Americans had won only the Indy 500 in the World Championship.
- Future world champions Jim Clark and John Surtees debut in Formula 1
- Alan Stacey and Chris Bristow die in the Belgium Grand Prix. The two British drivers died in separate incidents, very close to each other, and a few minutes apart.

- Ritchie Ginther drives the first rear engine F1 Ferrari at Monaco.
- Scoring changed to 8 (1st place), 6 (2nd), 4 (3rd), 3 (4th), 2 (5th) and 1 (6th). Points are no longer awarded for fastest lap.
- First win drivers: P. Hill, Rathmann
- First win constructors: Lotus
- Drivers from the following countries drove in this season: Britain, Argentina, Italy, France, Germany, Sweden, Australia, New Zealand, Venezuela, USA, Netherlands, Belgium, Portugal
- New nationalities represented: Venezuela
- Marques which raced for the first time: JBW, Scarab

1961

- First year of the 1.5-liter formula.
- Ferrari's shark nose cars dominate the season.
- Giancarlo Baghetti wins in his championship debut. He had previously won non-championship races.
- Stirling Moss' last season. He is the first driver to finish in the top 3 in the championship for seven straight seasons, yet never won the title.
- All cars that started the Dutch Grand Prix finished the race, and none pitted.
- Scoring changed again, for drivers: 9 (1st place), 6 (2nd), 4 (3rd), 3 (4th), 2 (5th) and 1 (6th). Scoring for constructors was not changed.
- Title favorite Wolfgang Von Trips dies at Monza.
- First American world champion, Phil Hill
- First win by a German driver.
- First win drivers: Von Trips, Baghetti, Ireland
- Drivers from the following countries drove in this season: Britain, USA, Germany, New Zealand, France, Sweden, Australia, Belgium, Switzerland,

South Africa, Netherlands, Italy, Argentina, Mexico, Canada
- New nationalities represented: South Africa, Mexico, Canada
- Marques which raced for the first time: De Tomaso, Gilby, Ferguson

Porsche had a brief existence as a works team in the early 1960s.

1962

- For the second year running a driver with the last name Hill wins the championship. Phil and Graham are not related, one is American, the other a Briton. Another Hill, Damon, would win the title in 1996...
- Brabham debuts as a constructor.
- The monocoque chassis is introduced in Formula One, in the Lotus 25
- First South African Grand Prix
- First win drivers: G. Hill, Clark, Gurney
- First win constructors: Porsche
- Drivers from the following countries drove in this season: Britain, USA, Italy, South Africa, Netherlands, Sweden, Mexico, Germany, New

Zealand, Australia, Belgium, France, Switzerland, Argentina, Rhodesia
- New nationalities represented: Rhodesia
- Marques which raced for the first time: Emeryson, Lola, ENB, Brabham

1963

- First Scottish world champion.
- First Mexican Grand Prix.
- Different times. Jim Clark used the same set of tires in the first four races of the season, winning three of them. He would just change them in the British Grand Prix.
- Jim Clark wins seven of ten races.
- First championship won by Lotus.
- First win drivers: Surtees
- Drivers from the following countries drove in this season: Britain, USA, New Zealand, South Africa, Sweden, Australia, Belgium, France, Switzerland, Netherlands, Italy, Germany, Portugal, Canada, Mexico, Rhodesia
- Marques which raced for the first time: ATS, BRP, Scirocco, Stebro, LDS, Alfa Special

1964

- John Surtees wins the World Championship due to questionable Ferrari teamwork, in Mexico. Rather than red, the Ferraris bear N.A.R.T.'s white and blue colors.
- Future world champion Jochen Rindt debuts in F1.
- Honda debuts in Formula 1.
- Carel Godin de Beaufort dies in the German Grand Prix.
- First Austrian Grand Prix.
- First win drivers: Bandini

- First win constructors: Brabham
- Drivers from the following countries drove in this season: Britain, USA, Sweden, Switzerland, Italy, France, Australia, New Zealand, Netherlands, South Africa, Belgium,. Germany, Austria. Portugal, Rhodesia, Mexico
- New nationalities represented: Austria
- Marques which raced for the first time: Derrington-Francis, Honda

Jim Clark features high up in everybody's list of all-time greats

1965

- First race of the season is held on January 1. The second is held only May 30.
- Future 3-time champion Jackie Stewart debuts in Formula 1.
- Jim Clark skips the Monaco race because he is busy elsewhere, winning the Indy 500… He thus

- becomes the first F1 driver to win the event. Jim would still win the 1965 title, plus six races.
- Except for Mexico, where Americans scored fastest lap and pole, all wins, fastest laps and poles were scored by British drivers.
- Future champion Hulme debuts in Formula 1.
- First win drivers: Stewart, Ginther
- First win constructors: Honda
- Drivers from the following countries drove in this season: Britain, New Zealand, Switzerland, Australia, South Africa, Italy, Austria, Rhodesia, USA, Sweden, Belgium, Germany, Mexico
- No French drivers raced in the championship for the first time.

1966

- Beginning of the 3.0-liter regulation.
- Many of the teams run 2.0 cars for part of the season.
- BRM introduces a H16 engine which does not perform well. BRM in fact runs the 2.0-liter engine for most of the season, and uses this engine to win Monaco (Stewart)
- John Surtees quits Ferrari mid-season, after winning a Grand Prix. He also manages another win driving for Cooper-Maserati and is the last driver to win races for two different constructors on the same season.
- John Taylor dies in Germany.
- Jack Brabham becomes the first driver to win the championship driving a self-built car. He wins four GPs in a row, in the 3.0 Repco engined car.
- First win drivers: Scarfiotti
- Drivers from the following countries drove in this season: Britain, Australia, Austria, New Zealand,

Italy, USA, France, Switzerland, Mexico, Belgium, Germany, Sweden
- Marques which raced for the first time: McLaren, Eagle, Matra

Cooper-Maserati was competitive in 1966, and its drivers finished second and third in the Championship, even though Surtees also drove for Ferrari.

1967

- First New Zealand world champion.
- Jim Clark wins four races in the new Cosworth engine car, but Denis Hulme wins the title on the strength of many second and third places, besides his two wins.
- First win by a Mexican driver.
- Dan Gurney becomes the second driver to win a F1 driving his self-built car. The American has a nice run, for the preceding week he won the 24 Hours of Le Mans sharing a Ford with A.J. Foyt.
- First Canadian Grand Prix.
- Lorenzo Bandini dies in the Monaco Grand Prix.
- First win drivers: Hulme, Rodriguez
- First win constructors: Eagle

- Drivers from the following countries drove in this season: New Zealand, Britain, Italy, Austria, Mexico, Belgium, USA, Australia, South Africa, Rhodesia, Switzerland, France, Sweden, Germany, Canada
- Marques which raced for the first time: Protos

The Eagle-Weslake that won in Belgium, 1967

1968

- So-called beginning of commercial sponsorship in Formula 1.
- Jim Clark beats Fangio's record for most wins in South Africa, but dies in a F-2 race at Hockenheim.
- First win by a Swiss driver.
- First win by a Belgian driver.
- Bruce McLaren wins the Belgian Grand Prix in a McLaren. It is the marque's first win.
- Wings begin appearing on Formula One cars.
- Jo Schlesser dies in the French Grand Prix, driving the new Honda.
- Future champion Mario Andretti debuts, scoring pole in the U.S. Grand Prix. He was not allowed to start in the Italian Grand Prix.
- First win drivers: Ickx, Siffert
- First win constructors: Matra, McLaren

- Drivers from the following countries drove in this season: Britain, Italy, Belgium, Austria, Australia, New Zealand, South Africa, Mexico, France, USA, Switzerland, Rhodesia, Sweden, Germany, Canada

Wings were introduced in Formula 1 in 1968. Pictured here a Ferrari

1969

- First win by an Austrian driver.
- Last time a Cooper races in F1.

- Gerhard Mitter dies in qualifying for the German Grand Prix.
- Only 13 cars start the French Grand Prix.
- First championship where all races were won by Cosworth powered cars
- First win drivers: Rindt
- No Italian drivers qualify even for a single race
- Drivers from the following countries drove in this season: Britain, Belgium, Austria, New Zealand, Australia, France, Germany, South Africa, Mexico, Canada, USA, Switzerland, Sweden, Rhodesia

1970

- First posthumous champion.
- First non-English speaking champion since 1957.
- First Austrian champion.
- March starts its F1 history on the right track, with three poles on the first four races and a win in the second. With a 2-car works team and quite a few privateer cars, the fields are boosted. However, March's success is not maintained during the season or during its history.
- First win by a Brazilian driver.
- Team owner Bruce McLaren dies testing a Can-Am car in Goodwood.
- Piers Courage dies in the Dutch Grand Prix
- Jochen Rindt dies in qualifying for the Italian Grand Prix.
- In his last season, Jack Brabham loses two Grand Prix to Jochen Rindt on the last lap.
- Spa-Francorchamps dropped as a Grand Prix venue.
- Alfa Romeo returns to Formula 1 as an engine supplier. The 8-cylinder engine is not successful in 1970 (McLaren) and 1971 (March).

- Future world champion Emerson Fittipaldi debuts in Formula 1
- First win drivers: Regazzoni, E. Fittipaldi
- First win constructors: March
- Drivers from the following countries drove in this season: Britain, Mexico, Italy, France, Brazil, South Africa, Germany, Switzerland, Belgium, New Zealand, Australia, Canada, Sweden, Austria, USA, Spain, Rhodesia
- Marques which raced for the first time: March, Bellasi, Tyrrell, Surtees

The March 701 with its typical sidepods was raced in 1970 and 1971. Here is Pescarolo in South Africa (Russell Whitworth)

1971

- The Tyrrell, which debuted in the end of the previous season, wins 7 of 11 races, and its drivers finish 1st (Stewart) and 3rd (Cevert).
- Ronnie Peterson is runner-up without winning a race.
- A turbine powered Lotus appears in some events but fails to impress. In its last entry, the car is entered under the banner of World Wide Racing, in golden colors, rather than the usual Team Lotus and Gold Leaf. The purpose was to avoid

impounding by Italian authorities, who were still investigating Rindt's death in 1970.
- Slick tires are introduced in Formula 1.
- Future world champion Niki Lauda debuts in F1.

- First win drivers: Andretti, Cevert, Gethin
- First win constructors: Tyrrell
- Drivers from the following countries drove in this season: Britain, France, Sweden, Brazil, USA, Mexico, Austria, Germany, New Zealand, Australia, South Africa, Rhodesia, Spain, Netherlands, Belgium, Switzerland, Italy, Canada

Jackie Stewart was unchallenged champion in 1971 (Rob Neuzel)

1972

- First Brazilian champion.
- Marlboro enters Formula 1 as a sponsor. The chosen team, BRM, fields as many as five cars, a large proportion of the field. At one point, there is speculation that as many as 8 BRMs would be

entered in races! BRM wins its last Championship race that season, at Monaco.
- Firestone champion for the last time.
- Future world champion Jody Scheckter debuts in Formula 1

Carlos Reutemann debuted in F1 in 1972, scoring pole in his first Grand Prix (Alejandro de Brito)

- After much expectation, Tecno enters Formula 1 with a 12-cylinder proprietary engine. The car is not a success.
- First win drivers: Beltoise

- Drivers from the following countries drove in this season: Britain, Brazil, Argentina, USA, Canada, France, Belgium, Switzerland, Italy, New Zealand, Australia, South Africa, Sweden, Rhodesia, Austria, Germany, Spain
- Marques which raced for the first time: Connew, Eifelland, Tecno Politoys

1973

- First Brazilian Grand Prix.
- First Swedish Grand Prix.
- Jackie Stewart beats Jim Clark's record for most wins

James Hunt debuted in 1973, driving a March for privateer team Hesketh.

- Future champion James Hunt debuts in F1.
- All races are won by Ford Cosworth equipped cars (for the second time) on Goodyear tires (first time).
- A pace car is used for the first time in the Canadian Grand Prix, and incorrectly deployed in front a back marker, establishing great confusion in timing and scoring.
- Jody Scheckter causes a multiple-car collision in the initial phases of the British Grand Prix, resulting in a red flag.
- The top 6 in the Monaco GP finish in the same order as the championship final ranking: Stewart-Emerson-Peterson-Cevert-Revson-Hulme.

- Gijs Van Lennep scores the first point for a Williams designed car, in Holland. Then in 1975, he would score Ensign's first point as well
- Roger Williamson dies in the Dutch Grand Prix
- Francois Cevert dies in qualifying for the U.S. Grand Prix.
- First win drivers: Peterson, Revson
- Drivers from the following countries drove in this season: Britain, Brazil, Argentina, USA, France, Belgium, Switzerland, Italy, New Zealand, Australia, South Africa, Denmark, Sweden, Netherlands, Liechtenstein, Austria, Germany
- New nationalities represented: Liechtenstein, Denmark
- Marques which raced for the first time: Iso Marlboro, Shadow, Ensign

1974

- First win by a South African driver.
- First win by a driver from Africa.
- A large number of new constructors enter Formula 1: Hesketh, Penske, Parnelli, Amon, Maki, Lyncar, Trojan and Token. None of them stay active on the long run, but Hesketh and Penske do win a race apiece within the next two seasons.
- Peter Revson dies testing a Shadow in Kyalami.
- Even though most motorsport suffered from poor economic conditions worldwide, F1 prospered and besides the established and new teams, many privateers enter races during the season.
- Helmut Koinigg dies in the U.S. Grand Prix at Watkins Glen.
- First win drivers: J. Scheckter, Lauda, Reutemann
- Drivers from the following countries drove in this season: Brazil, Argentina, USA, France, Belgium, Switzerland, Britain, Italy, New Zealand, Australia,

South Africa, Denmark, Sweden, Netherlands, Liechtenstein, Finland, Austria, Germany, Canada
- New nationalities represented: Finland
- Marques which raced for the first time: Hesketh, Amon, Token, Trojan, Penske, Parnelli, Maki, Lyncar

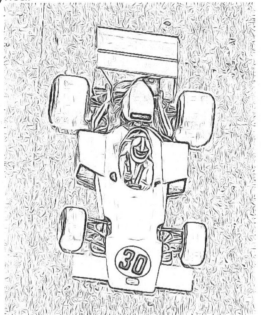

A sketch of one of several versions of the 1974 Amon Formula 1 that raced only in that season

1975

- Half points assigned to drivers for the first time, in the interrupted Spanish Grand Prix, due to an accident involving the leader at the time, Rolf Stommelen.
- The Austrian Grand Prix is also interrupted due to excessive rain, and half points are awarded for the second time during the season.
- Firestone leaves F-1 after the Argentine Grand Prix
- A female driver scores for the first time: Lella Lombardi, 0.5 point in Spain

- Future World Champion Alan Jones debuts in Formula 1
- Mark Donohue dies from injuries sustained in the Austrian GP practice.
- Graham Hill and other team personnel die from an airplane crash during the off-season.
- First time a South American constructor participates in Formula 1, Fittipaldi (Argentina)
- First win drivers: Pace, V. Brambilla, Hunt, Mass
- First win constructors: Hesketh
- Drivers from the following countries drove in this season: Brazil, Argentina, USA, France, Belgium, Switzerland, Britain, Italy, New Zealand, Australia, South Africa, Sweden, Netherlands, Japan, Austria, Germany, Ireland
- New nationalities represented: Japan
- Marques which raced for the first time: Fittipaldi, Hill, Williams, Stanley-BRM

Hesketh's single GP win, Holland 1975 (Pieter Kamp)

1976

- First race in Asia.

- First Japanese Grand Prix
- Dunlop and Bridgestone tires used in the Japanese Grand Prix.
- First U.S. West race held at Long Beach, California. The U.S. becomes the first country to officially have two F1 races in every season.
- Last Formula 1 run at Nürburgring's Nordschleife.
- Last time a locally entered South African car participates in the South African Championship (Ian Scheckter, Tyrrell)
- First time two female drivers attempt to qualify for a race, in the British Grand Prix: Lella Lombardi and Divina Galica. Neither qualifies.
- The Tyrrell team debuts a 6-wheel car, the P34. The car scores many points, the drivers end the championship 3rd and 4th, but it is obviously slower than McLaren and Ferrari.
- Brabham adopts Alfa Romeo 12-cylinder engines.
- First win drivers: Watson
- First win constructors: Penske

The 6-wheel Tyrrell P-34 raced in the 1976 and 1977 seasons.
(Alejandro de Brito)

- Drivers from the following countries drove in this season: Brazil, Argentina, USA, France, Belgium, Switzerland, Britain, Italy, New Zealand, Australia, South Africa, Denmark, Sweden, Netherlands, Austria, Japan, Germany, Spain, Ireland
- Marques which raced for the first time: Boro, Kojima, Ligier

1977

Wolf did well in its debut season, then faded (Kurt Oblinger)

- Wolf wins its debut race in Argentina, then goes on to win two more, and claim second in the driver's championship.
- A turbocharged car is entered in the championship for the first time (Renault), in Britain. The turbocharger breaks.
- The 12-cylinder Matra engine finally wins its first Championship race, in Sweden, 9 years after its introduction.
- Lotus introduces the wing car to F1 and wins five races during the season.

- Michelin enters Formula 1 as a tire supplier.
- Tom Pryce dies from a freak and avoidable accident in Kyalami.
- Carlos Pace dies from a small plane crash in native Brazil, after a particularly good beginning of the season.
- First win drivers: A. Jones, Nilsson, Laffite
- First win constructors: Wolf, Shadow, Ligier
- Drivers from the following countries drove in this season: Austria, South Africa, Argentina, USA. Brazil, Mexico, Netherlands, Belgium, Australia, Germany, Britain, France, Sweden, Sweden, Japan, Switzerland, Italy, Spain, Finland, Canada
- Marques which raced for the first time: Wolf, Appolon, Lec, McGuire, Renault

1978

- First win by a Canadian driver.
- Lotus 79 introduces the concept of ground effects in F1 and becomes the class of the field.
- A couple of future champions debut in F1, Nelson Piquet and Keke Rosberg.
- Ronnie Peterson dies as a result of injuries sustained in the Italian Grand Prix.
- First points by a turbo car.
- First win drivers: G. Villeneuve, Depailler
- Drivers from the following countries drove in this season: Austria, South Africa, Argentina, Canada, Australia, Switzerland, Brazil, Finland, Sweden, Italy, France, USA, Britain, Germany, Mexico, Ireland, Belgium, Spain, Netherlands
- Marques which raced for the first time: Merzario, ATS, Arrows, Theodore, Martini

Renault's boldness soon paid off (Alejandro de Brito)

1979

- Improved Lotus 79 copies, the Ligier and Williams, win many races during the season.
- Williams wins its first race, in Britain, surprisingly driven by Clay Regazzoni.
- Alfa Romeo returns as a constructor in Formula 1.
- Niki Lauda again quits a team before the end of the season, and this time retires. For a little while, at least.
- Brabham shifts back to Ford Cosworth engines at the end of the season.
- First win by a turbo car, French Grand Prix (Renault)
- First win drivers: Jabouille
- First win constructors: Williams, Renault
- Drivers from the following countries drove in this season: South Africa, Britain, Canada, Austria, Australia, Brazil, Switzerland, Mexico, Belgium, Argentina, France, USA, Netherlands, Italy, Germany, Ireland, Finland

- Marques which raced for the first time: Kauhsen

1980

- Williams wins its first championship, Alan Jones.
- Popular driver Regazzoni has a career ending accident at Long Beach.
- McLaren begins using carbon fiber in Formula 1.
- After finishing 1-2 in 1979, Ferrari drivers have a poor showing in the Championship, and Jody Scheckter even fails to qualify for a race in Canada.
- 1978 champion Mario Andretti also does badly, scoring one point from 6th in the US Grand Prix, and leaves Lotus at the end of the season.
- Future world champions Nigel Mansell and Alain Prost debut in Formula 1.
- The Spanish Grand Prix, won by Alan Jones, is stripped of official Championship status, due to the ongoing fighting between teams and authorities for control of the sport. Ferrari, Renault and Alfa Romeo boycott the race.
- South African Desire Wilson is entered in a privateer Williams in the British Grand Prix but fails to qualify. She is the only female driver to attempt qualification in the 80s.
- Patrick Depailler dies in a testing accident for Alfa Romeo.
- First win drivers: Piquet, Pironi, Arnoux
- Drivers from the following countries drove in this season: Britain, Australia, Brazil, France, Argentina, Canada, Finland, Switzerland, Mexico, Italy, South Africa, Ireland, USA, Netherlands, Sweden, Germany, Austria, New Zealand
- Marques which raced for the first time: Osella

1981

- Ferrari adopts turbo engines and becomes the second turbo engine car to win in the championship at Monaco and Spain.
- This is the first championship run under the name FIA Formula One Championship, instead of World Championship for Drivers
- First San Marino Grand Prix, and Italy becomes the second country to have two races in the same season.
- Watkins Glen is replaced by Las Vegas as the site of the US Grand Prix, for obvious reasons, no longer called East. Long Beach remains in the calendar.

The Brabham BT 49 was around between 1979 to 1982, and won the 1981 championship (Alejandro de Brito)

- The double chassis Lotus 88 is banned. Designers keep on looking for ways to circumvent regulations, or interpret them creatively.

- A C version of the Brabham BT49, uses hydropneumatic suspension, designed to circumvent ride height during motion.
- FISA and FOCA fighting means the South African Grand Prix is stripped of its Championship Status. The British teams attend, while French and Italian teams, aligned with FISA, fail to go.
- Alan Jones retires from F1.
- First win drivers: Prost
- Drivers from the following countries drove in this season: Britain, Australia, Brazil, France, Argentina, Finland, Switzerland, Mexico, Canada, USA, Ireland, Italy, Netherlands, Chile, Sweden, Spain
- New nationalities represented: Chile
- Marques which raced for the first time: Toleman

1982

- Carlos Reutemann decides to retire from F1, so Williams has a driver problem.
- First Finnish world champion.
- First win by a Finnish driver.
- In a very dramatic season, no less than 11 drivers driving for 7 constructors win races.
- BMW becomes the third turbo engine to win in F1, in Canada.
- Niki Lauda returns to F1 and wins two races in his comeback season.
- Riccardo Paletti dies from injuries in the Canadian Grand Prix.
- Gilles Villeneuve dies from an accident in practice at Zolder, Belgium.
- The San Marino Grand Prix is boycotted by British teams.
- A third US race is held at Detroit for the first time.

- A Swiss Grand Prix is held at Dijon, France, so multiple races are held in the same season in 3 countries for the first time.
- First win drivers: Rosberg, Patrese, De Angelis, Alboreto, Tambay
- Drivers from the following countries drove in this season: Britain, France, Argentina, Austria, Canada, Sweden, Germany, Italy, Ireland, Chile, Brazil, Colombia, USA, Switzerland, Netherlands, Spain, Finland
- New nationalities represented: Colombia

1983

- Nelson Piquet becomes the first driver to win the championship driving a turbo (BMW) car. By the end of the season, both Williams (Honda) and McLaren (TAG) are running turbo engines, so there are turbo offerings from Renault, Ferrari, Alfa Romeo, Hart, TAG and Honda.
- Porsche returns to Formula 1, as an engine supplier. The engine is badged TAG.
- Last time a normally aspirated car wins until 1989, Alboreto in Detroit, in a Tyrrell. This is the original Cosworth's engine 155th and last win. Different Cosworth designs would win races.
- A European Grand Prix is run at Brands Hatch, so two GPs ran in British territory on the same season.
- No first-time winning drivers and constructors. A pattern emerges where it becomes increasingly difficult for new winners to break into F1, both drivers and constructors.
- Drivers from the following countries drove in this season: Britain, Brazil, Finland, Austria, France, Switzerland, USA, Italy, Venezuela, Chile,

Colombia, Australia, Germany, Belgium, Canada, Sweden, Ireland.
- Marques which raced for the first time: Spirit

1984

- McLaren, with its TAG-Porsche engines, win 12 races, and Lauda takes the title by half a point over teammate Prost.
- Nelson Piquet gets pole for most events in his BMW engined Brabham but has scant luck in races.
- Ayrton Senna debuts in Formula 1, driving a Toleman-Hart.
- The Honda turbo engine wins for the first time.
- Long Beach is out of the calendar, and a race is held on the streets of Dallas, in addition to Detroit. Under scorching heat, the race takes a little over 2 hours to finish, and parts of the paving come off. 13 cars retire from accidents and the futile exercise is not repeated.
- All races are won by turbo engine cars for the first time.
- Alboreto becomes first Italian driver since Scarfiotti (1966) to win a race for Ferrari in Formula 1.
- For the second season in a row, there are no first-time winning drivers in a season.
- Tyrrell's points, and its drivers are excluded from the Championship, because Brundle's car is found to carry illegal fuel. To many, including, of course, Tyrrell, the punishment seemed excessive and caused immense losses to the constructor. The decision was appealed but upheld, and some felt that FIA was trying to woo more manufacturers to come with turbo engines into F1. No one was impressed by the stunt.

- Jo Gartner in an Osella and Gerhard Berger in an ATS finish 5th and 6th in the Italian Grand Prix, but get no points because their teams entered only one car in the championship, and both drivers raced second entries!
- Drivers from the following countries drove in this season: Britain, France, Finland, Italy, USA, Belgium, Switzerland, Austria, Brazil, Venezuela, Germany, Brazil, Netherlands, New Zealand, Sweden
- Marques which raced for the first time: RAM

1985

- Senna wins his first race in Portugal, under torrential rain.
- The Brazilian quickly shows talent for posting pole positions, achieving seven during the season.
- Renault quits as a works team at the end of the season but continues to supply engines to Lotus and Ligier for 1986.
- Williams-Honda proves to be the fastest car at the end of the season, winning the last three races
- First race held in Oceania.
- First Australian Grand Prix.
- First win drivers: Senna, Mansell
- Drivers from the following countries drove in this season: France, Italy, Sweden, Britain, Belgium, Germany, Austria, Brazil, USA, Finland, Netherlands, Australia, Switzerland
- Marques which raced for the first time: Minardi, Zakspeed

1986

- First Hungarian Grand Prix held. This is significant, being the first GP held behind the Iron Curtain, even before the demise of the Soviet Union.
- Turbo engines become mandatory.
- Williams-Honda wins a large number of races, but the failure to define who is the number one driver practically gives the title to Alain Prost in the McLaren-TAG.
- Curiously, this is the season in which Piquet won more races, four. In his three championship winning seasons he only won three.
- Toleman becomes Benetton. The cars are equipped with BMW engines.
- Popular Jacques Laffite has a bad accident in the British Grand Prix which ended his F1 career. Two races before he had finished second at Detroit.
- At the time of his retirement Jacques had become the driver with most GP starts ever, 176. Currently, 28 drivers have raced more times in F1, many of them drivers with no wins.
- Elio di Angelis dies testing a Brabham at Paul Ricard, a particularly low-drag body design. This is the last F1 death for a while, until Ratzenberger's in 1994.
- First win drivers: Berger
- First win constructors: Benetton
- Drivers from the following countries drove in this season: Brazil, Britain, France, Austria, Italy, Belgium, Germany, Sweden, Switzerland, Australia, Finland, Netherlands, Canada
- Marques which raced for the first time: Benetton, AGS

1987

- Even though Nigel Mansell won six races to Piquet's three, the Brazilian was more consistent (including seven 2nd places!) and won his third title.
- Nigel Mansell is out of the last two races of the season, due to injury.
- Lotus gets Honda engines.
- Alain Prost beats Jackie Stewart's record for most wins, held since 1973.
- FIA does not seem to make up its mind. First it does not want normally aspirated engines, then makes turbos mandatory, then it wants the normally aspirated engines back! The Jim Clark Cup and Colin Chapman Trophy are instituted for 3.5 liter normally aspirated cars, which will become the rule in 1989. Just a few teams show interest, and Tyrrell's Jonathan Palmer easily wins it. At Hockenheim the team has the best overall performance, 4th (Streiff) and 5th (Palmer).

Williams Honda won a single driver's title in Formula 1, 1987

- Unfair rule: Yannick Dalmas finishes 5th in Australia but does not get points because he is not a regular driver in the Championship.

- Drivers from the following countries drove in this season: Britain, France, Brazil, Sweden, Austria, Belgium, Japan, Italy, Germany, USA, Spain, Switzerland

1988

- The season is a McLaren show, as Senna and Prost win all but one race of the season.
- Just a few teams ran turbos (McLaren, Lotus, Arrows, Ferrari, Zakspeed, Osella) while most of the rest of the field runs 3.5 Cosworths, and a few, 3.5 Judds. All races are won by turbos.
- A few new teams appear, such as Coloni, Rial, EuroBrun, the re-born March and Scuderia Italia, and pre-qualifying is required as a rule for the first time.

The McLaren-Hondas were not only effective, they also had very smooth and pretty lines (Philip Morris)

- Mansell leaves Williams and is signed by Ferrari.
- Drivers from the following countries drove in this season: France, Austria, Brazil, Britain, Italy, Japan, Belgium, USA, Sweden, Spain, Argentina, Germany

- Marques which raced for the first time: Eurobrun, Rial, Dallara, Coloni

1989

- Mansell wins his first race for Ferrari, but things do not go all that well the rest of the year, as McLaren-Honda still dominates.
- Renault comes back to Formula 1, now with a 3.5 normally aspirated engine.
- Philippe Streiff crashes in pre-season testing at Jacarepagua, Brazil, and becomes quadriplegic and wheelchair bound.
- Turbo engines no longer allowed.
- Ayrton Senna is disqualified from a win in the Japanese Grand Prix, due to an on track altercation with Prost, and Alessandro Nannini is promoted to first.
- Prost is declared champion in Japan, as it becomes mathematically impossible for Senna to outscore the Frenchman after the disqualification.
- Senna accuses FIA President Jean Marie Balestre of favoring Prost.
- Phoenix replaces Detroit as the site of the single US Grand Prix.
- First win drivers: Boutsen, Nannini
- Drivers from the following countries drove in this season: Britain, France, Brazil, Italy, Japan, Germany, USA, Belgium, Switzerland, Sweden, Spain, Finland, Argentina, Austria
- Marques which raced for the first time: Onyx

1990

The Tyrrell 019 of 1990 was the first F1 car with a raised "dihedral" nose cone which became the norm in F1 design for many years.

- Ferrari has both Mansell and Prost on the team, and surprisingly, things go relatively smooth. Prost does win five races and finishes runner-up to Senna, while Mansell wins a single race and seems a bit lost.
- Nelson Piquet sensationally wins the two last races of the year for Benetton, his first wins since late 1987. Until then he only had a couple of podiums during the season.

Nelson Piquet did really well to win the last two races of 1990 in the Benetton

- Martin Donnelly has a horrifying crash at Spain and is lucky to survive.
- Senna and Prost again crash in Japan. Senna had pole but was unhappy with the dirty side in which he would start and asked Balestre to change it and was refused. Senna then deliberately crashes into Prost who got an advantage at the start.

- Drivers from the following countries drove in this season: Brazil, France, Belgium, Italy, Japan, Germany, Britain, Austria, Switzerland, Sweden, Finland, Australia
- Marques which raced for the first time: Leyton House, Life, Monteverdi

1991

- Mansell is back at Williams and finishes second in the championship.
- Alain Prost is booted out of Ferrari, after he refers to the cars as trucks, and decides to take a sabbatical year in 1992.
- Future world champion Mika Häkkinen debuts in Formula 1.

Ayrton Senna won his final title in 1991

- Scoring changed again: 10 (1st place), 6 (2nd), 4 (3rd), 3 (4th), 2 (5th) and 1 (6th).
- The Australian Grand Prix is stopped in the early stages of the race, due to torrential downpour, and half points are awarded. Until Belgium, 2021, it was the shortest World Championship event ever held.
- Drivers from the following countries drove in this season: Brazil, France, Italy, Japan, Britain, Finland, Austria, Sweden, Portugal, Germany
- Marques which raced for the first time: Jordan, Fomet, Footwork, Lambo

1992

- Nigel Mansell utterly dominates the championship in his Williams-Renault, finally winning it. Williams does not re-sign him for 1993 so he goes to the USA to drive in CART.
- Phoenix is dropped from the calendar, so for the first time there is not a single World Championship event in the USA.
- Brabham is permanently out of Formula One.
- Future world champion Damon Hill debuts in Formula 1
- In his first full season, Schumacher finishes third in the Championship.
- Giovanna Amati is the last woman to attempt to qualify for a F1 race. She drove for Brabham,
- Honda stepping out.
- First win drivers: M. Schumacher
- Drivers from the following countries drove in this season: Britain, Italy, Brazil, Germany, Austria, France, Finland, Japan, Belgium, Switzerland, Netherlands
- Marques which raced for the first time: Fondmetal, Andrea Moda, Venturi

1993

- Prost comes back, driving for Williams and wins his last title. Then he retires at the end of the season.
- After trying out the Lamborghini engines in the off-season, McLaren decides to run Peugeot engines.
- Honda engines are sold to customers as Mugen, starting an engine rebranding pattern in F1.
- Damon Hill sports the number zero in his Williams.
  - Ayrton Senna wins Monaco for the sixth time, beating Graham Hill's record.

Alain Prost, pictured here in his early F1 days at Renault, won his final title and retired in 1993, and for a few years was the winningest GP winner. (Alejandro de Brito)

- First win drivers: D. Hill
- Drivers from the following countries drove in this season: Britain, Brazil, France, Finland, Austria, Italy, Germany, Japan, USA, Portugal, Belgium
- Marques which raced for the first time: Sauber, Larrousse

1994

- First German champion.
- For the first time since 1986 there is a death in Formula 1. In fact, a death in a race had not happened since Paletti's death in 1982. Making matters worse, two drivers die at Imola on the second weekend, one of them Ayrton Senna, widely reckoned the best of his generation. The other deceased driver was Austrian Rolland Ratzemberger.
- Williams relies on Damon Hill to be the standard bearer for the team, and David Coulthard as number two for most races. Surprisingly, it hires Mansell yet again for a few races and a massive retainer, and Mansell does win for a last time.
- The Pacific Grand Prix is added to the calendar, in essence, a second race in Japan.
- The European Grand Prix is held in Jerez for the first time, so it is Spain's turn to have two races on the same season.
- Peugeot enters the championship as an engine supplier. The engine is used by McLaren for one season.
- The end of the original Lotus team.
- Mercedes officially reenters the World Champion as an engine supplier to Sauber.
- Drivers from the following countries drove in this season: Germany, Britain, France, Brazil, Japan,

Austria, Italy, Portugal, Australia, Netherlands, Monaco, Belgium, Finland, Switzerland
- Marques which raced for the first time: Simtek, Pacific

1995

- Benetton shifts to Renault engines and wins the Championship again.
- Schumacher/Hill finish 1-2 in the Championship for the second straight year.
- McLaren shifts to Mercedes engines, starting a long partnership.
- First win drivers: Coulthard, Alesi, Herbert
- Drivers from the following countries drove in this season: Germany, Britain, Austria, Finland, France, Japan, Italy, Brazil, Netherlands, Switzerland, Portugal, Denmark, Belgium
- Marques which raced for the first time: Forti

The early days of Red Bull sponsorship, the Sauber in 1995, driven by Frentzen (Alejandro de Brito)

1996

- Damon Hill finally wins a title, after finishing second to Schumacher for two straight seasons. At the end of the season he is not re-signed by Williams.
- Marlboro ends its association with McLaren dating back to 1974. From that point on Marlboro would support only Ferrari in F1.
- One-hour shootout qualifying adopted, which lasted until 2002. On Saturday drivers would have 12 laps to set their fastest time, which would decide Sunday's grid.
- Future world champion Jacques Villeneuve debuts in Formula 1
- First win drivers: J. Villeneuve, Panis
- Drivers from the following countries drove in this season: Britain, Canada, Austria, Finland, France, Germany, Brazil, Japan, Portugal, Italy, Netherlands

1997

- Williams hires Heinz Harald Frentzen for Damon Hill's place, but the German proves disappointing, winning a single race.
- First Canadian champion.
- Adrian Newey joins McLaren as a designer and team performance improves substantially.
- Michael Schumacher's points are taken away from him, due to his misdeeds in the Canadian Grand Prix, where he was deemed to have crashed onto Jacques Villeneuve on purpose, fighting for the championship.
- Bridgestone joins Goodyear as a regular tire supplier.
- First Mercedes engine win since 1955

- Alain Prost buys Ligier. His young team leads in Austria, the car driven by a young Jarno Trulli.
- First Luxembourg Grand Prix, held at Nürburgring, Germany
- First win drivers: Häkkinen, Frentzen
- Drivers from the following countries drove in this season: Britain, Germany, Finland, Austria, France, Italy, Japan, Brazil, Denmark, Netherlands, Canada, Argentina
- Marques which raced for the first time: Stewart, Prost

1998

Mika Häkkinen won the first of his title in 1998 in commanding style
(Alejandro de Brito)

- Renault is officially out of Formula 1, and its engines become the Mecachrome and Playlife. Neither Williams nor Benetton win a single race.
- First championship by a driver of a Mercedes powered car since 1955
- Goodyear withdraws from Formula 1 at the end of the season.
- First win constructors: Jordan

- Drivers from the following countries drove in this season: Britain, Finland, Brazil, Germany, Italy, France, Japan, Argentina, Netherlands, Denmark, Austria, Canada

1999

- First Malaysian Grand Prix. Beginning of shift to addition to Asian races in detriment of European events.
- All cars use Bridgestone tires for the first time.
- First win drivers: Irvine
- First win constructors: Stewart
- Drivers from the following countries drove in this season: Finland, Britain, Italy, Germany, France, Japan, Austria, Spain, Canada, Brazil
- Marques which raced for the first time: BAR

Jordan was in winning form in 1999.

2000

- Future world champion Jenson Button debuts in Formula 1
- Ferrari wins the driver's championship for the first time since 1979.

- Stewart becomes Jaguar.
- V10 engines become mandatory, the idea being to keep costs down.
- BMW returns to F1 as engine supplier to Williams
- First win drivers: Barrichello
- Drivers from the following countries drove in this season: Britain, Germany, Finland, Brazil, France, Spain, Netherlands, Argentina, Canada, Italy, Austria
- Marques which raced for the first time: Jaguar

2001

- Future world champions Kimi Räikkönen and Fernando Alonso debut in Formula 1
- Michael Schumacher beats Alain Prost's record for most wins
- Mika Häkkinen announces he would take a sabbatical at the end of the season but never returns
- Michelin returns as tire supplier.
- First win by a Colombian driver
- First win drivers: Montoya, R. Schumacher
- Drivers from the following countries drove in this season: Britain, Germany, Brazil, Finland, Colombia, Italy, France, Canada, Spain, Argentina, Czech Republic, Netherlands, Malaysia
- New nationalities represented: Czech Republic, Malaysia

2002

- Toyota debuts as a Formula 1 team, built from scratch. Operation is based in Germany.
- The Japanese team scores only two points, the same score as Minardi and Arrows.

Michael Schumacher won 5 straight titles between 2000 to 2004, one of the few records Lewis Hamilton was unable to match (Alejandro de Brito)

- Arrows does not survive the season, and its final race was in Germany. The team had been around since 1978, and never won a race.
- Ferrari wins a total of 15 races: 11 wins for Schumacher and 4 wins for Barrichello, and the drivers finish 1-2 in the championship.

- Third placed driver Montoya does not win a single race and scores a little more than a third of the points scored by the champion.
- Drivers from the following countries drove in this season: Britain, Spain, Germany, Finland, Brazil, Colombia, Italy, Japan, Canada, Australia, Malaysia, France
- Marques which raced for the first time: Toyota

2003

- An extremely consistent Räikkönen finishes runner-up to Michael Schumacher, even though he won a single race during the season.
- A total of 8 drivers win races during the year, representing five teams (Ferrari, McLaren, Williams, Renault and Jordan).
- The Brazilian Grand Prix is ended in the 56th lap, because of two track blocking crashes. The win is initially awarded to Räikkönen, is then awarded to Fisichella in the Jordan, which had appealed the original result.
- Cristiano da Matta surprisingly leads 17 laps of the British Grand Prix, in the Toyota.
- One-lap qualifying adopted and kept with some changes for three seasons.
- Scoring changed again, and points are given to the top 8: 10 (1st place), 8 (2nd), 6 (3rd), 5 (4th), 4 (5th), 3 (6th), 2 (7th) and 1 (6th).
- First win by a Spanish driver
- First win drivers: Räikkönen, Alonso, Fisichella
- Drivers from the following countries drove in this season: Britain, Ireland, Spain, Germany, Finland, Brazil, Colombia, Italy, Hungary, Australia, Canada, Japan, Denmark, Netherlands, France.
- New nationalities represented: Hungary

2004

- Jenson Button finishes third in the Championship, without a single win.
- The first Grand Prix is held in the Middle East.
- Michael Schumacher beats his won record for most driver wins in a season, with 13 victories.
- As teammate Barrichello won two races, Ferrari again wins 15 races, as other teams win merely 3 races.
- The drivers placed 3rd and 4th in the championship do so without wins.

A sketch of the all-conquering 2004 Ferrari F2004

- First Chinese Grand Prix.
- First Bahrain Grand Prix.
- First win drivers: Trulli
- Drivers from the following countries drove in this season: Britain, Spain, Germany, Finland, Brazil, Colombia, Italy, Canada, Japan, Australia, Austria, France, Hungary

2005

- Jaguar is sold to Red Bull.

- Renault finally wins a Championship as a Formula 1 team/constructor, after 28 years.
- After five years of domination, Ferrari finally gives some sign of weakness, as Michael Schumacher won only a single race during the season, the poorly supported U.S. Grand Prix.
- Kimi Räikkonen scores ten fastest laps during the season.
- As Michelin tires prove dangerous in the Indianapolis track, cars equipped with that brand are withdrawn from the race, and only Bridgestone shod cars start. As a result, only six cars start, the smallest grid even in F-1, which gives a chance to both Jordans and Minardis to score points.
- Last points by Minardi, which is sold to Red Bull at the end of the season and becomes Toro Rosso in 2006.
- First Spanish world champion
- First Turkish Grand Prix
- Drivers from the following countries drove in this season: Britain, Spain, Germany, Finland, Brazil, Japan, Italy, Australia, Colombia, Austria, Canada, Portugal, India, Netherlands
- New nationalities represented: India
- Marques which raced for the first time: Red Bull

2006

- BAR Honda becomes Honda. The team wins a race during the season.
- Michael Schumacher retires at the end of the season, winning seven races and finishing second in the championship.
- Elimination is adopted in qualifying for the first time, and slightly tweaked until stabilizing in the current format in 2010.

- 2.4-liter V8 engines introduced. Dispensation was given to Toro Rosso to continue using the 3.0-liter engine, with rev limiter.
- Sauber becomes BMW-Sauber and Williams loses the BMW engine contract
- Adrian Newey joins Red Bull.
- First Red Bull podium in Monaco (Coulthard). The car raced with Ferrari engine.
- American Scott Speed races for Toro Rosso but fails to score.
- After winning ten races in 2006, McLaren wins nothing in 2006.
- First win drivers: Massa, Button
- Drivers from the following countries drove in this season: Britain, Spain, Germany, Finland, Brazil, Italy, Colombia, Australia, Austria, Netherlands, Canada, Poland, Portugal, USA, France, Japan
- New nationalities represented: Poland
- Marques which raced for the first time: Toro Rosso, Midland, Super Aguri

2007

- An intrateam battle finishes badly for McLaren. After Hamilton and Alonso battle each other all year long, Ferrari's Kimi Räikkönen gets 3 wins in the last four races and takes the Championship from the pair.
- Future world champions Lewis Hamilton and Sebastian Vettel debut in Formula 1
- McLaren's points are taken away in the Constructor's Championship, due to Spygate. A McLaren engineer improperly received secrets from a Ferrari mechanic, and someone had to take the blame. The driver's points are retained.

- As a result, BMW Sauber is promoted to runner-up in the Constructors championship, although it did not win a single race.
- Ferrari uses explicit tobacco advertising for the last time in the Chinese Grand Prix. From then on the team would still be sponsored by Marlboro, using several means of subliminal advertising, including the Mission Winow brand. The team would still be called Scuderia Ferrari Marlboro until halfway through the 2011 season.
- First win drivers: Hamilton
- Drivers from the following countries drove in this season: Britain, Spain, Finland, Brazil, Germany, Italy, Poland, Australia, Austria, Japan, USA, Netherlands
- Marques which raced for the first time: Spyker

2008

- Despite winning his home race in Brazil, getting pole and the fastest lap, Brazilian Felipe Massa loses the championship in the last lap as Lewis Hamilton overtakes German Timo Glock and finishes 5th. That is enough to give the Brit an advantage and his first title.
- Another year, another scandal. At Singapore, Nelsinho Piquet allegedly crashes his car voluntarily, at the behest of authorities in the Renault team, to favor teammate Fernando Alonso, who ends up winning the race. The information only comes to light after Piquet is fired from the team in 2009.
- Second-string Toro Rosso actually wins a race before main team Red Bull.
- The Singaporean Grand Prix is the first night race in Formula 1.
- First European Grand Prix at Valencia, Spain.
- First Singaporean Grand Prix

- First win drivers: Vettel, Kubica, Kovalainen
- First win constructors: Toro Rosso, BMW-Sauber
- Drivers from the following countries drove in this season: Britain, Spain, Finland, Brazil, Germany, Poland, Japan, Australia, Italy, France
- Marques which raced for the first time: Force India

2009

Sebastian Vettel won four straight championships between 2010 to 2013, but did not sustain that form at Ferrari.

- Brawn, a team hastily put together for the 2009 season from the ashes of the withdrawing Honda team, with little visible sponsorship, wins the championship and eight GPs in the process. That turns out to be the only season contested by the team, which becomes Mercedes in 2010.
- Red Bull finally becomes a top Formula One team, winning six races, including the last three of the season.

- BMW departs from F1, but Sauber would continue to be known as BMW Sauber in 2010. The team was bought back by Peter Sauber at reasonable terms.
- Last season of competition for Toyota.
- First Abu Dhabi Grand Prix
- First win drivers: Webber
- First win constructors: Brawn, Red Bull
- Drivers from the following countries drove in this season: Britain, Spain, Germany, Brazil, Finland, Italy, Poland, France, Japan, Switzerland, Australia
- Marques which raced for the first time: Brawn

2010

- In 2009, four new teams had been approved to join Formula 1 during the season to boost up the grids somewhat. One, USF1 never even makes to the track. The performance of Hispania, Virgin and Tony Fernandes' Lotus leaves a lot to be desired.
- KERS units banned.
- The current qualifying format, consisting of Q3, Q2 and Q1, with elimination, is adopted this season.
- First Red Bull championship title.
- Adrian Newey can now claim his designs won championships for three teams, Williams, McLaren and Red Bull.
- Mid-race refueling banned.
- First South Korean Grand Prix
- Return of Mercedes to Formula 1, having acquired the Brawn Grand Prix team.
- Return of Michael Schumacher to F-1, in what turned out to be a 3-year stint.
- Drivers from the following countries drove in this season: Britain, Spain, Germany, Brazil, Australia,

Poland, Russia, Switzerland, Finland, India, Austria, Japan, India
- New nationalities represented: Russia
- Marques which raced for the first time: Virgin, HRT

2011

- A legal battle emerges for the right to use the Lotus name in F-1. The Malaysian team headed by Tony Fernandes believes it has such right, but Lotus Cars, with forges a partnership with Renault, claims rights to use the brand and logo. A British Court rules in favor of Lotus Cars.
- KERS units become optional.
- DRS (drag reduction system) is introduced
- New scoring system adopted: points given to the top 10: 25 (1st place), 18 (2nd), 15 (3rd), 12 (4th), 10 (5th), 8 (6th), 6 (7th), 4 (8th), 2 (9th) and 1 (10th).
- Pirelli becomes the sole tire supplier in F1.
- First Indian Grand Prix
- Drivers from the following countries drove in this season: Britain, Germany, Brazil, Australia, Spain, Russian, Venezuela, Japan, Mexico, Switzerland, India, Italy, Belgium

2012

- Fernandes' Lotus becomes Caterham. Renault becomes Lotus GP.
- Return of Kimi Räikkonen to F1, driving for Lotus GP. The Finn signs a contract with a low fixed retainer and a performance clause, where he would be paid EUR50,000 for each point scored. The driver ends up third in the Championship, and

- scores more than 200 points, earning a substantial amount of money.
- Formula 1 returns to the United States, race held at Austin.

Pastor Maldonado surprised everybody with a win in 2012, never to be repeated. Williams has also not won a race since then. (Unattributed photo)

- Mercedes' first win as a works team upon return since 1955 (Rosberg)
- Michael Schumacher retires permanently from F-1, after a 3-season stint at Mercedes.
- First win by Venezuelan driver
- First win drivers: Maldonado, N. Rosberg
- First win constructors: Lotus GP
- No Italian drivers race in the Championship for the first time.
- Drivers from the following countries drove in this season: Britain, Germany, Brazil, Finland, Australia, Spain, France, Japan, Mexico, Venezuela, Russia, India
- Marques which raced for the first time: Caterham, Marussia

2013

- Sebastian Vettel matches Michael Schumacher's record for most wins in a season, 13. He also wins the last nine races of the season.
- Lewis Hamilton moves over to the Mercedes camp and wins a race. Teammate Nico Rosberg wins two.
- Kimi Räikkonen again scores a load of points (183), earning EUR50,000 per point, driving Lotus to desperation. He scores six second places plus one win and does not race in the last two races of the season.
- Mark Webber finishes 3rd in his swansong season in Formula 1.
- Drivers from the following countries drove in this season: Britain, Finland, Germany, Mexico, Australia, Brazil, France, Venezuela, Netherlands, Spain

2014

- Kimi Räikkönen returns to Ferrari
- For the first time one of the three teams introduced to F1 in 2010 scores points (Bianchi driving the Marussia at Monaco)
- After the awesome season in 2013, Vettel does not win a single GP, while new teammate, Australian Daniel Ricciardo, wins three.
- Jules Bianchi suffers a very serious accident in the Japanese Grand Prix.
- First Mercedes title since 1955
- First Russian Grand Prix
- First win drivers: Ricciardo
- Drivers from the following countries drove in this season: Britain, Finland, Germany, Mexico,

Australia, Sweden, Japan, France, Venezuela, USA, Denmark, Russia, Spain, Brazil

Lewis Hamilton began his domination of F1 in 2014 (Mercedes-Benz)

2015

- Jules Bianchi becomes the first driver to die from injuries sustained in a F1 race since Ayrton Senna in 1994. He was hospitalized for many months due to injuries suffered in the 2014 Japanese Grand Prix. He crashed at a whopping 254G, the worst ever recorded in a F1 crash
- Sebastian Vettel's first season at Ferrari. The German won in his second start for the team, in total, he had 3 wins and finishes 3rd in the Championship.
- Only Mercedes and Ferrari win races.
- Future world champion Max Verstappen debuts in Formula 1, at 17 years of age, driving for Toro Rosso. He scores points in 10 races and finishes 12th in the championship.

- McLaren finishes the Constructors Cup in 9th, outscoring only Marussia, that had 0 points. It is just the beginning of the team's most difficult period.
- The Virtual Safety Car is adopted.
- Drivers from the following countries drove in this season: Britain, Finland, Germany, Mexico, Australia, Spain, Venezuela, Denmark, Russia, Sweden, Brazil, Netherlands, France

2016

- Champion Nico Rosberg shockingly retires at the end of the season.
- Max Verstappen wins his first race for Red Bull, in Spain.
- Max finishes all but four races in the points and is fifth in the Championship.
- Only Mercedes and Red Bull win races, and the best Ferrari can do is five 2nd places.
- Haas debuts in F1 and sensationally scores in the first race.
- First win by a Dutch driver.
- First win drivers: M. Verstappen
- Drivers from the following countries drove in this season: Britain, Finland, Germany, Mexico, Australia, Netherlands, France, Spain, Belgium, Indonesia, Russia, Denmark, Sweden, Brazil
- New nationalities represented: Indonesia
- Marques which raced for the first time: Haas, Manor

2017

- First Azerbaijan Grand Prix
- After 40 long years Bernie Ecclestone leaves as CEO of Formula One Group, as the American

- company Liberty Media acquires the sport. Ross Brawn is named new CEO.
- Toro Rosso shuffles its drivers often during the season, and Sainz, Kyvat, Gasly and Hartley all drive the cars at one point or another.
- Force India finishes fourth in the Constructors Cup as both Sergio Perez and Esteban Ocon score in most races. The best placing in a race is 4th, however.
- On the other hand, McLaren outscores only Sauber, and finishes ninth, repeating the poor 2015 performance.
- Drivers from the following countries drove in this season: Britain, Finland, Germany, Mexico, Australia, Netherlands, Denmark, France, Belgium, Spain, Sweden, Italy, Brazil, Russia, Canada, New Zealand

2018

- The halo protective device is introduced to protect drivers' heads in impact situations. Cars become uglier, but drivers are safer. Plus there is an additional place to put sponsor's names…
- There are no driver changes during the season, and all participating drivers score in the championship, the first time this happens.
- Force India changes ownership in the middle of the year, and the points for the original owner are excluded.
- As a result of this exclusion Haas finishes fifth in the Constructors' Cup.
- McLaren adopts Renault engines and its fortune changes.
- Alfa Romeo returns to Formula 1, as branding for the Sauber team.
- No Brazilian driver races in the Championship for the first time since 1970.

- The championship becomes more balanced as Red Bull and Ferrari manage to win, together, 10 GPs.
- Sebastian Vettel manages to keep the pressure on Lewis Hamilton for most of the Championship, but starting in Italy the Brit's performance improves greatly.
- Toro Rosso signs with Honda as engine supplier.
- Drivers from the following countries drove in this season: Britain, Finland, Germany, Mexico, Australia, Netherlands, France, Denmark, Belgium, Spain, Sweden, New Zealand, Canada, Russia, Monaco

2019

- New Ferrari signing Leclerc wins two races on a row and outscores Vettel in the championship.
- Leclerc also gets seven pole positions during the season.
- Kevin Magnussen gets the fastest lap in the Singapore Grand Prix, driving a Haas.
- Fastest lap is again awarded 1 point.
- After the McLaren woes of recent years, 2019 is Williams' turn to hit bottom. The team finishes dead last in the Constructors' Cup, scoring a single point.
- First win by a Monegasque driver.
- First win drivers: Leclerc.
- Drivers from the following countries drove in this season: Britain, Netherlands, Mexico, Australia, Canada, Italy, Finland, Germany, Monaco, Denmark, France, Spain, Thailand, Russia, Poland
- Marques which raced for the first time: Racing Point

2020

- Formula 1 navigates relatively well the height of the Covid-19 pandemic. Rather than insisting on enforcing contracts, one-off deals are done with several tracks, including three Grand Prix in Italy for the first time and two in Austria. Many races are cancelled, but 17 events are ultimately held.
- As a result, a number of races appear in the calendar for the first time (some for the last time): the Tuscan, 70th Anniversary, Styrian, Eifel, Emilia Romagna and Sakhir Grand Prix.
- Lewis Hamilton matches Michael Schumacher record of 7 titles.
- Sergio Perez wins the Sakhir Grand Prix in sensational fashion, after falling to last in the first lap of the race.
- Williams fails to score a single point in the Championship.
- One of its driver's, George Russell, substitutes Hamilton at Mercedes, who has Covid-19, in the Sakhir Grand Prix. George immediately leads and almost wins the race.
- Haas does slightly better than Williams, scoring 3 points.
- First win drivers: Perez, Gasly
- First win constructors: Racing Point
- Drivers from the following countries drove in this season: Britain, Spain, Mexico, Netherlands, Monaco, Italy, France, Canada, Australia, Thailand, Brazil, Denmark, Germany, Finland, Russia
- Marques which raced for the first time: Alpha Tauri

2021

- A controversial decision is taken to stop the stormy Belgian Grand Prix after a single lap behind a safety car and award half points to the top 10 drivers. At the end of the season these points make a huge difference.
- The first ever sprint qualifying race is held at Silverstone, on a Saturday. The new format allows drivers a chance to score a few more points than normal (3 for 1st, 2 for 2nd and 1 for 3rd). The result of the race determines the grid for the full-points Sunday race.
- Another controversy occurs in the final race of the year, as, after a late race Safety Car Intervention, race director decides to remove the yellow flag for the final lap, essentially giving an advantage to Max Verstappen, who had brand new tires, as opposed to Hamilton, who had old. The Dutch driver passes Hamilton and wins the Championship.
- First Dutch F1 champion (Max Verstappen)
- First Saudi Grand Prix.
- First Qatar Grand Prix. There are now four Grand Prix in the Middle East.
- Lewis Hamilton becomes the first driver to obtain more than 100 wins in F-1.
- Renault becomes Alpine.
- Racing Point becomes Aston Martin.
- First win drivers: Ocon.
- First win constructors: Alpine.
- Drivers from the following countries drove in this season: Britain, Spain, Mexico, Netherlands, Monaco, France, Canada, Australia, Poland, Italy, Germany, Russia, Japan, Finland
- Marques which raced for the first time: Alpine

2022

- In the off-season there was great speculation that Lewis Hamilton may not return to F1, due to the outcome of the 2021 championship, but he does come back, after all.
- The engines used by Red Bull and Alpha Tauri are now known as Red Bull, as the company bought the engine program from Honda.
- The first Chinese driver in Formula 1, Zhuo Ghanyu has a very scary crash in the beginning of the British Grand Prix, but his life is likely saved by the halo.
- In the wake of the Russian war against the Ukraine, Haas terminates its contract with Russian sponsor Uralkali and driver Nikita Mazepin. As a result, Haas calls back Kevin Magnussen.
- For the first time since the early 2010s Mercedes struggles, and Lewis Hamilton, particularly, has trouble making the car go fast. Mercedes cars are affected by porpoising (excessive bouncing) more than other top cars, which resulted from new aerodynamic ground effect regulations, devised to increase competition.
- The latest generation Mercedes engines also seem to be slower than the latest Red Bull and Ferrari offerings.
- Scoring for the Saturday sprint races changes. Now the first eight get points, as follows: 1st (8 pts), 2nd (7 pts), 3rd (6 pts), 4th (5 pts), 5th (4 pts), 6th (3 pts), 7th (2 pts), 8th (1 pt.)
- Miami Grand Prix run for the first time.
- First win drivers: Carlos Sainz and George Russell
- Max Verstappen wins the championship for the second year running, in commanding style.
- In the process Max matches, then passes both Sebastian Vettel and Michael Schumacher as the

winningest driver in a single season, earning a total of fifteen victories.
- First, and surprising, pole position for Haas for the Braziian sprint racel.
- Vettel retires at the end of the season, and Alonso is hired by Aston Martin.

Red Bull is the newest dominating team in Formula One (Red Bull)

- There is quite a bit of controversy involving the hiring of Oscar Piastri as a McLaren driver, for the Australian had a contract with Alpine.
- 8-race winner Daniel Ricciardo to sit out the 2023 season; however, he is to be paid a large retainer to be Mercedes' reserve driver.
- Generally driver moves are common at the end of a season. In 2022, there was an atypical number of changes in team principals. Mattia Binotto left Ferrari, who hired Alfa's Fred Vasseur. Afla Romeo replaced Vasseur with McLaren's Andreas Seidl. Seidl, on the other hand, was replaced by Andre Stella. Jost Capito left Williams!
- Drivers from the following countries drove in this season: Britain, France, Spain, Netherlands, Japan,

China, Mexico, Australia, Canada, Finland, Thailand, Monaco, Germany, Denmark
- New nationalities represented: China

George Russell is one of the most promising younger drivers (Mercedes-Benz)

2023

- A couple of drivers debut in the beginning of the season, American Logan Sargeant and Australian Oscar Piastri.
- Logan is the first American in F1 since Alexander Rossi's exploits of 2015. He quickly becomes crash prone, not seeming to make much progress during the season.
- Piastri, who replaced another Australian at McLaren manages to finish 8th at his home race, led a lap, started on the front row, posted a fastest lap in Italy, got several podiums at the time of writing, and more importantly, won the sprint race in Qatar, in fact winning for McLaren before highly rated Lando Norris. He looks like a star of the future.

- The driver he replaced, Daniel Riccardo, returned to F1 at Alpha Tauri. Daniel took the place of Nick De Vries who underperformed, but was sidelined after just two races, due to a crash. He is replaced by Liam Lawson, from New Zealand, who quickly displays speed and scores in his third race.
- As for veterans, the season started very well for Fernando Alonso, whose move to Aston Martin was seen with some doubt by pundits.
- Alonso broke some of his own records, scoring three straight podiums in the first three Grand Prix, scored in the first 14 races, finished second a couple of times, led on occasion, became the first driver to complete 100,000 km in F1, and first to complete 20,000 laps in the sport.
- Alonso also got his 100th podium in Saudi Arabia, after his penalty is removed.
- Aston Martim thus led for the first time.
- The Emilia Romagna Grand Prix is cancelled due to extreme rains in Italy.
- Red Bull wins the first 14 races of the season, breaking McLaren's 1988 record. Perez won two of these races, to Verstappen's 12 at the time.
- As in 1988, it is Ferrari who spoiled the party, as Sainz wins the Singapore Grand Prix.
- Miami is a true Latin paradise, and the top 3 qualifiers in the year's race were Spanish speakers.
- In qualifying some of the slowest teams displayed surprising form. In Canada Albon finished Q2 in first place, driving a Williams. Tsunoda was P1 in Q1 in Monaco and in Singapore while Zhou Ghanyu scored top time in Q1 in Hungary. Haas also surprised on occasion.
- In Holland there were four different leaders in the first few laps and Alex Albon managed to stay on soft tires for an amazing 45 laps.

- Behind Verstappen, the season is quite interesting. Besides the quick Aston Martin, McLaren's fortunes are also revived, while Ferrari's performance also improved. Mercedes still looked far from a win but Hamilton did score his 9th pole in Hungary and Alpine squeezes a couple of podiums.
- Felipe Massa announced a lawsuit in an effort to cancel the 2008 Singapore Grand Prix and earn the year's title. The chances of success are seemed with skepticism and causes outrage in some corners.
- FIA finally approved the Andretti-Cadillac entry in Formula 1, but the other constructors still have to approve the addition of an eleventh team, which is unlikely to happen for financial reasons: no one wants to make less money.
- That would be General Motors' first F1 effort.
- As for Porsche's future entry in F1, it seemed to have fizzled
- The results of sprint races no longer determine the grid for the Sunday race, and there is a separate qualification for the sprint race.
- Several penalties handed down for off-track antics in Austria, and Esteban Ocon is punished with a whopping 30 seconds!
- Helmut Marko, the Red Bull team advisor, makes off color remarks concerning team driver Sergio Perez' heritage. He is widely criticized for it, but at the time of writing was not booted out of F1.
- Although a night race, the Qatar Grand Prix was held in extreme hot conditions, leading drivers to vomit inside their helmets (Ocon) and others to report losing conscience at the track and passing out after the race.
- Verstappen wins his 3rd title in Qatar. With 5 races to go, he has 14 wins under his belt for the season.

GRIDS

Nowadays we are used to standard 2-2 grids in all Formula 1 races, but that was not so back in the early days of the sport.

In fact, the grid for the first World Championship race, in **Silverstone**, 1950, was 4-3-4, which was common in the 50s. Times were different: cars and tires were narrow, and safety considerations few, for drivers and spectators.

An even busier grid was used in **Monza**'s early races, 4-4, but, as you see below, that quickly changed.

The last 4-3-4 grid was used at the **Nürburgring**, in the 1967 German Grand Prix, which had separate grids for Formula 1 and Formula 2 cars.

The 3-3 grid was used at the **Indy 500** in all races valid for the World Championship, from 1950 to 1960, and was used a few times in **Monza**.

A popular grid formation was 3-2-3, but that lasted until 1973. The last Grand Prix with this formation was the Dutch Grand Prix, but other races featured that grid formation that season, such as Argentina, Brazil, South Africa, and Britain. As of the 1973 Dutch Grand Prix, all grids had a 2-2 formation, either same line, straight grid (one car per line) or staggered grid (with less spacing between cars).

Local promoters and clubs had quite a bit of discretion establishing regulations and practices at the time, in fact, entry lists were directly negotiated by teams, drivers and promoters. So they could choose whatever grid formation they wished. It seems that **Monza** could not make their mind as to what was the best grid formation for its track.

As said, in the early races, **Monza** used a 4-4 grid formation. In 1953-54, the Italians used a 3-3 grid formation, which changed to 3-2-3 in 1956, and then 4-3-4. In 1960 **Monza** had a 3-2-3 formation, and in the full grid of 1961, a 2-2 formation was used which did not prevent the accident that killed **Von Trips** right at the start. 3-2-3 was used in 1964 to 1968, and the organizers finally settled on the 2-2 grid starting in 1969.

We are also used to stable entry lists these days, where teams basically enter the same drivers and numbers of cars race-in, race-out. The situation was very different, the farther back you go. Organizers could accept entries as they pleased, as long as the maximum engine displacement limit was observed. Thus, in the second race of 1950, **Harry Schell** was entered in a 1.1 JAP 2-cylinder, engined **Cooper** Formula 3 car at **Monaco**. In years ahead, Formula 2 cars were often entered, even a midget in a U.S. Grand Prix, **Porsche** sports cars, streamlined (wheel covered) Formula 1s that looked more like sports cars, while old pre-war **ERA** voiturettes with supercharged engines were also entered in 1950. The idea was to provide fuller grids, for even though works teams would enter as many as 5 cars on occasion, there were few such works teams around. Privateers were necessary, even though some of them were not that skillful, and their cars questionably kept. Cars for local drivers were often sent to the host countries for hire.

That meant that the first time two consecutive Formula 1 races had the same drivers and cars on the grid was in 1974, in Argentina and Brazil. This only happened because **Rikki Von Opel** had handling issues in his **Ensign** in practice and did not take part in the Argentine race and did not even go to Brazil. The only reason that was achieved in that pair of races was that attending the South American races was very expensive for teams, given the distance to Europe, and no privateers would come. In 1973 **Surtees** provided a third car to local driver **Luis Bueno**, and in 1975 **Jarier** failed to start Argentina, thus the grids were different. That situation was far from stabilized in 1974, for that year a large number of new teams debuted, and quite a few privateer teams raced **Brabhams, Marches, Surtees**, etc. Additionally, some of the poorer works teams often rented rides to a number of drivers. There were surprises in every race, and many drivers did not qualify for races. Eventually, things became more organized and bureaucratic, and you could basically memorize the entry list that would apply all season, except for injury, disease or eventual driver firing or team bankruptcy. It should also be noted that most F1 drivers drove in other series until the late 70s, and injuries and even death often caused the change of team make-up.

Most races in 1974 had 25 cars on the grid, including Monaco, where the busy grid resulted in the elimination of a few cars right at the beginning of the race. From that point on, **Monaco** would have only 20 starters. The race at **Nivelles**, Belgium, had 31 cars on the grid. Kinnunen in a privateer Surtees failed to qualify, so potentially, 32 cars could have started the race.

The Indy 500 already had the traditional 33 starters during the 50s, so it was usually the race with the largest number of starters during the 1950 to 1960 seasons. However, that was not the largest grid ever. The 1952/1953 German Grand Prix at the **Nürburgring** welcomed large number of entries of German Formula 2 cars such as the **Veritas, BMW, EMW** and **AFM**, thus, the 1953 race had a whopping 34 cars on the grid. Other races with large grids were **Monza,** 1961 (32 cars, one car, driven by **Andre Pilette**, did not qualify) and Britain, 1952 (32 cars), while the smallest grid was the U.S. Grand Prix of 2005, where only six cars took the start owing to safety concerns involving Michelin tires that shod most of the entries – only Bridgestone tired cars participated.

As said, entry lists were very busy in 1974, but also in most races of 1977. That often led to pre-qualification sessions, which would eliminate the slowest cars, which were often entered by privateer teams. In the late 80s, entry lists grew to almost unmanageable numbers, as quite a few **Cosworth** engined constructors appeared in competition. As you can see in the table below, the numbers dwindled in a few years:

Season	Usual number of entries
1988	31
1989	39
1990	35
1991	34
1992	32
1993	26
1994	28
1995	26
1996	22/20
1997	22

CAR NUMBERS

F1 car numbers are now associated with drivers, who choose a number they will use permanently in the series. This has been common practice in NASCAR for decades, allowing drivers to be branded and affiliated with a number, such as **Richard Petty**, who was forever linked to the number 43, and **Dale Earnhardt** to number 3. A favorite combination is the use of repeated numbers in two digits, options chosen by **Lewis Hamilton** (44), **Max Verstappen** (33) and **Carlos Sainz Junior** (55). For the 2022 season Max is using number 1, a prerogative of the past season champion, just to show who is boss.

Nowadays drivers "own" numbers of their choice. George Russell chose 63 (Mercedes Benz)

Organizers were not bound to any numbering restrictions until the 70s, in fact they could assign numbers as they pleased. This state of affairs persisted until 1972, a season in which eventual world champion **Emerson Fittipaldi** used many numbers. For 1973, the World Champion would be entitled to use the number 1 all season, and his teammate, number 2. Thus Emerson had number 1, and **Ronnie Peterson**, number 2. The rule did not last long, as Stewart became champion in 1973 but retired at the end of the season. The solution was to let **Lotus**, the Constructor's Cup Champion, continue to use numbers 1 (Peterson) and 2 (**Ickx**).

The largest number ever used in the Championship was 208, used by **Lella Lombardi** in her attempt to qualify a Brabham for the 1974 British Grand Prix. Lella was sponsored by a Radio Station (Luxembourg), whose frequency was 208. Lella did not qualify, but was 28th out of 34 drivers.

The Germans always came up with something different in the 50s, and cars for the 1952 German Grand Prix were numbered from 101 to 135. The highest number of the lot fell to **Ernst Klodwig**, in a **BMW Heck** Formula 2 car.

As for the lowest number, 0 was used a few times. **Jody Scheckter** used the number a couple of times in the third **McLaren** in 1973, no reason given, while **Damon Hill** used the number in the 1993 and 1994 seasons because **Williams** had been champion in both 1992 (**Mansell**) and 1993 (**Prost**), however, the champions did not drive in the team or elsewhere in F1 in the next season. Nigel was not retained by Williams for 1993, while Prost retired for good at the end of that season. In keeping with the tradition, Damon had number 0 both seasons, and in fact, managed to win quite a few races with it. No negative numbers have been used.

Former skier **Divina Galica** was not superstitious and was entered in a number 13 **Surtees** for the 1976 British Grand Prix. Challenging the unlucky number did no good, and Divina was not fast enough to qualify for the race. Next time round she used a 24.

SPONSORS

Most people would wrongly say that the first commercial sponsorship appeared on cars in the World Championship for Drivers in the 1968 Spanish Grand Prix, in a **Lotus**. That sounds true, would make great copy, except that the **Indy 500** was part of the Championship from 1950 to 1960, and a large number of the cars in the American race were named and carried stickers from sponsors, in fact, some of them became famous household names, such as **Bowes Seals Fast, Dean Van Lines, Blue Crown**, etc.

Additionally, the **BRP** team of the early 60s sneakily promoted their patrons **Yeoman Credit Bank** (a car financing company) in the 1960 season. Dressed in a beautiful and distinctive light green color, the partnership lasted a single season, as the company shifted support to **Reg Parnell's** team for the 1961 season. **UDT Laystall**, a Yeoman competitor, was signed up as BRP's sponsor. By naming the teams after the sponsors, these companies got good enough of a return for their troubles, mostly displaying the cars in their print advertising.

A further detail that is easy to escape us, is the fact that one race before the 1968 Spanish Grand Prix, the South African Grand Prix, **John Love** appeared driving a car in the colors of sponsor Gunston, a South African cigarette brand. So the Gold Leaf urban legend is only that, an urban legend.

Outside of that, yes, commercial sponsorship started in earnest in 1968, and it has remained the lifeblood of Formula 1 teams: without sponsors Formula 1 is unlikely to have survived and been relevant to this day.

That first sponsor was a cigarette brand, **Gunston**, followed by **Gold Leaf**, and in fact, tobacco invested billions of dollars in the sport until being disallowed altogether. Cigarette makers sponsored many of the champions from 1968 until funding Ferrari in concealed fashion in the 2000s. **Marlboro**, which started sponsoring teams in 1972, won many races and championships, and often sponsored more than one team (plus many drivers) during a given season.

Even the famous Paris Moulin Rouge night club sponsored a Formula 1 entry (Kurt Oblinger)

Cigarette brands from France, Japan, South Africa, Brazil, Austria, Finland, Germany, U.S., Portugal, Britain and others were featured on F1 cars: **Gitanes, Camel, Benson & Hedges, Colt, Lucky Strike, Hollywood, JPS, Memphis, Gunston, Rothmans, West, Chesterfield, 555, Galouise, L&M, Lexington, Embassy, Lark, Mild 7, Viceroy, Gigante, Winfield, Barclay,** among others.

A large number of industries invested in F1 in the course of the decades, and the required investment was puny compared to the current millions, which means that the profile of sponsors has changed substantially with time.

For instance, in the 70s, model car manufacturers such as **Politoys, Matchbox, Norev** and **John Day** appeared as name sponsors in some Formula 1 entries. While **Schuco** in the 2000s and **Tamiya** in the 1990s also had their logos splashed on F1 cars, they were not name sponsors: sometimes you needed a microscope to find the sticker on the car!

Perhaps the most unusual sponsor appearing on a F1 car was a funeral home that sponsored the **Merzario** team for a while in 1979. The spooky sponsorship was made even spookier, given the racer's black and yellow color scheme.

Sponsors with some questionable connections, such as **Leyton House, Moneytron** and **Ambrosio**, also financed F1 efforts. Enough said about that.

Automakers also appeared as sponsors in teams with no clear technical partnership. Most famously, **Aston Martin** appeared as Red Bull's sponsor recently, which before carried **Infiniti (Nissan)** stickers while the relationship with Renault prospered. Truck manufacturers **Volvo, Leyland, Iveco, Magirus** and **Scania** also joined the frame for a while, and so did Russia's **Lada** and India's **Tata Motors**.

The long-arm of fraud has financed F1 projects, such as the Moneytron Ponzi scheme that sponsored Onyx.

Another unusual sponsor is **ABBA**, the Swedish musical group, which sponsored **Slim Borgudd** for a while. In addition to being a driver, Slim was a session musician who played with ABBA on occasion.

Bin Laden is also an unusual sponsor. No, we are not talking about Osama. **Bin Laden** happened to be the largest construction company in Saudi Arabia, and it was one of the Saudi companies that sponsored **Williams** in the early 80s, along with **Saudia Airlines, TAG** and **Albilad**.

Besides the above, a large breadth of industries have sponsored Formula 1 teams and drivers since 1968: credit card, commercial bank, investment bank, watches, soft drink, beer, hard liquor, wine, automobile magazines, sugar, oil companies, toiletries, cosmetics, perfumes, alarm systems, skis, snowboards, generators, news magazines, airlines, property company, pharmaceuticals, coffee, photocopiers, energy drinks, clothing, computer, printers, electronics, tiles, government agencies, mineral water, phone operator, chemicals, appliances, tea, sound equipment, cigar, water technology, tools, real estate developer, wheels, temporary job agencies, websites, men's magazine, preservatives, rolling paper, oil additives, credit bureau, software, underwear, dental cleaning equipment, aviation maintenance companies, cookware, car rental, foods (salt, cooking oil, dairy products, cheese, candy, chewing gum, ice cream, spices, processed meats, gastronomic delicacies, sea food), bicycles, fire extinguishers, night club, chewing tobacco, refrigeration equipment, movies, lighting systems, business consulting, loud speakers, auditing, insurance, video games, athletic wear, steel, hotel, sunglasses, auto parts, paint, musical instruments, tv stations, airplanes, pens, computer chips, lottery, heavy equipment, natural gas, electronics component distribution, crypto currency, furniture, shaving goods, motorcycles, heating equipment, apps, recreational vehicles, art gallery, carpets, home cleaning products, fitness equipment, utilities meters,toothpaste, charities, fertilizer, car dealerships, vacuum cleaners, online casinos, regular casinos, lubricants, tortillas, ball valves, cybersecurity, even a private group of stockbrokers. Superman and a Ponzi scheme fraud, both of which amount to the same thing.

SOME YPF CURIOSITIES

YPF is the Argentine oil company. It was the first state owned oil company outside the Soviet Union, and remained so until 1999, when it was privatized and bought by Repsol. As Argentina keeps on shifting between right- and left-wing governments, it was nationalized again in 2012.

The small YPF decal on the 1955 Mercedes is visible just because there was nothing else on the car... (Alejandro de Brito)

YPF has also paid many of the bills for Argentine car racing while a state owned company, and, as you can see on the 1955 Mercedes driven by Fangio, it was not shy placing its decals on Grand Prix participants.

Upon return of the Grand Prix in the oficial calendar, 1972, YPF went one step further. As the sponsor of the race, YPF felt entitled to place its logo on all cars on the field, despite the fact that the only car that actually had a fuel contract with YPF was Reutemann's car. So other fuel companies, such as Esso, Shell, Fina, BP, Texaco had to stomach seeing the logo of a competitor placed on cars sponsored by them. This was particularly noticeable in the Tyrrells, which were sponsored by French oil company Elf.

Stewart in Argentina, 1972. Double oil company sponsorship! (Alejandro de Brito)

In the end, no harm was done. Most people outside of Argentina had no idea what YPF meant, for the company operated locally, whereas all other mentioned oil companies had international operations.

In these days of multitudes of lawyers, accountants, business advisors and managers one would never see a situation like this.

ENGINES

Constructors that participated in the world championship for a while and did not build their own engines had their cars equipped with various powerplants. **Ferrari**, the longest standing team/manufacturer in F1 has mostly used its own engines, although a **Jaguar** engined Ferrari chassis appeared in the 1950 Italian GP, driven by **Clemente Biondetti**, and **Lancia** powered cars were used in the 1956 and 1957 seasons by the Scuderia.

McLaren, which has been around since 1966, has used engines from many manufacturers. Until settling with the **Cosworth** engine in 1968, **Bruce McLaren** used **Ford, Serenissima** and **BRM** engines in the 1966 and 1967 seasons, none of which proved suitable. In 1970, 8-cylinder **Alfa Romeo** engined cars were fielded for Italians **Andrea de Adamich** and **Nanni Galli**, without success. The first turbo engine used by **McLaren** was the **TAG-Porsche** engine, which powered the cars from 1983 to 1987, resulting in 3 straight championships (1984 to 1986). That was replaced by the **Honda** in 1988, which stayed on board until 1992. After one season using **Ford Cosworth** engines, McLaren cars were equipped with **Peugeot** engines, a partnership which did not gel. **Mercedes** engines were used starting in 1995, resulting in the last championships won by McLaren. The partnership with Honda from 2015 to 2017 did not bring the expected results, and **Renault** engines also proved to be disappointing. McLaren is again using Mercedes engines.

The traditional **Lotus** team used **Climax** engines from the start until 1966, but a rare **Maserati** engined **Lotus** was used by **Prince Starabba** in the 1961 Italian Grand Prix. **BRM** powered Lotus appeared more often, fielded by the **Parnell** team in the early 60s, and then by the works in the 1966 season. The 16-cylinder engine was not fast enough, even though **Clark**'s genius was enough to give it a victory in the 1966 U.S. GP. The Cosworth engine was initially used by Lotus in 1967, and right away the partnership resulted in a pole (**G. Hill**) and win (Clark) in the 1967 Dutch Grand Prix. Except for the **Pratt and Whitney** turbine used in the 56 Lotus of 1971, in three races, Lotus cars used the Cosworth engine exclusively until the turbo era, when a partnership was forged with **Renault**, in 1984. In spite of **Senna**'s talent, the Lotus-Renault won only a few races, and by 1987 the cars were equipped with **Hondas** for two seasons. The next engine used by the constructor was the **Judd** engine, in 1989 and 1991, which was not a winning combination. In the latter years of the team, **Lamborghini** engines were used without success in 1990, then replaced by Cosworth (92,93) and **Mugen Honda** engines in 1994. **Ford** Formula 2 engines were also used in Lotus cars that appeared in the German Grand Prix Formula 2 section.

Cooper was around from 1950 to 1969, but it did use a variety of power plants during its period in the championship. The first one was a 1.1 **JAP** twin motorcycle 2-cylinder engine used by **Harry Schell** in a Formula 3 Cooper in the 1950 **Monaco** Grand Prix, which proved slow. Harry was eliminated by accident, at any rate. The **Bristol** engine was used in the Formula 2 years, and occasionally afterwards, but most of Cooper's success was achieved with **Climax** engines, including the majority of Cooper's wins in Formula 1. A **Borgward** engine was tried briefly in 1959 in the Formula 2 class, and 2.5 **Maserati** engines from the venerable 250F were used in the last years of the 2.5 liter Formula. A 1.5 Maserati was briefly used in the early years of the 1.5 liter regulation, while a 3.0 Maserati achieved Cooper's last wins in F-1 in 1966 and 1967. **Osca** and **Ferrari** engines were used in the 2.5 liter era, while a 1.5 liter **Alfa Romeo** engine was used in a Cooper in South Africa. Cooper used **BRM** engines in the 1968 season, achieving podiums in Spain (**Redman**) and

Monaco (**Bianchi**). The last Cooper outing, in 1969, was a car equipped with a Maserati engine.

Brabham also used quite a few power plants during the years. In the 1.5-liter era Brabham used **Climax** engines in his works cars, although **BRM** engined cars appeared in the hands of privateers. The **Repco** engine was adopted in 1966, winning the 1966 and 1967 championships, however, by 1968 it proved to be lacking in power. Brabham shifted to **Cosworth** in 1969, staying with the power plant until a tie-up with **Alfa Romeo** in 1976. The Alfa partnership only lasted until late 1979, and before the season was over Brabham resumed use of the **Cosworth**, winning the 1981 championship. The turbo partner was **BMW**, with which Brabham won the 1983 championship. In the last Brabham years in F-1, already out of **Ecclestone**'s hands, Brabham used both **Yamaha** and **Judd** engines. **Ford** engined Formula 2 Brabhams also appeared in the German Championship's F2 section.

Brabham used a number of engines from 1962 to 1992, including Alfa Romeo. This is the BT-46, driven by Lauda in 1978. (Kurt Oblinger)

CARS WITH LONG CAREERS

In the age of hundreds of millions of dollars budgets, teams introduce new models yearly, and updates every other race. Formula 1 was not flush with money for a great part of its history, so a few successful cars were used by a number of constructors until they were no longer competitive.

A sketch of the venerable Maserati 250F

I suppose the most famous one is the **Maserati 250F**. Quite a few of these cars were produced and sold to privateers in Europe (Britain and the Continent), South America, USA and Australia/New Zealand. Some of these cars were fitted with Chevrolet engines and raced in Formula Libre way into the 60s, but in the World Championship they were always equipped with the 2.5 Maserati 6- and 12-cylinder engines. In addition to selling cars to privateers, Maserati ran a works team until 1957, and won races in **Fangio**'s and **Moss**' hands, plus dozens of races in several guises driven by many drivers outside the World Championship. Fangio won in the car's debut in

1954, and after taking another win in Belgium, went over to Mercedes. The 250F was Fangio's last mount in the 1958 French Grand Prix, but by then the front engined car was no longer competitive. Notwithstanding, some 250Fs still appeared as late as the 1960 Argentine Grand Prix, while the engines were fitted in more nimble Cooper chassis. A new generation Maserati never used by the works was sold and raced as the **Tec-Mec** by **Fritz D'Orey** in the 1960 U.S. Grand Prix. The 12-cylinder engine initially developed for the 250F turned out in the 3.0-liter era **Coopers** starting in 1966.

The **Lotus 49** stuck around in Formula 1 between 1967 and 1971. The car was the mount used to debut the **Cosworth** DFV engine, in the Dutch Grand Prix of 1967, winning and scoring pole right off the bat. **Clark** would still win three other races in 1967, plus the 1968 South African Grand Prix, then died in a F-2 race at Hockenheim. His teammate at Lotus **Graham Hill** carried the torch well and won the 1968 championship. The car continued its winning ways in 1969, driven by Hill and **Rindt**, but it was no match for the new Matra. Lotus continued to use the car in early 1970 races, in fact, Rindt won Monaco driving a 49C. The 49 continued to be used by the Lotus works, the Walker and the **Pete Lovely** team in 1970, and was also sold to South African teams, winning many races there. The last Lotus 49 participations in the World Championship took place in 1971, in the form of a Lotus 49/69 put together by American **Pete Lovely**, which raced in the North American races but it was no longer competitive. The works did use the type in non-championship races in 1971, driven by **Wilson Fittipaldi Junior** and **Tony Trimmer**. Curiously, a number of World Champions drove the 49 at one time or another: **Jim Clark, Graham Hill, Jochen Rindt, Emerson Fittipaldi** and **Mario Andretti**.

The Lotus 72 raced from 1970 until 1975, and won two championships, in the hands of **Jochen Rindt** and

Emerson Fittipaldi. Rindt won four straight races with the 72 in 1970, building a sizeable championship cushion that would not be overcome, even though he did not race in the last four events. Fittipaldi proved the car was excellent by winning his fourth start in Formula 1 in 1970. After a disappointing season in 1971, Fittipaldi became the class of the field and won the 1972 title in commanding style. By then, 72s had been sold in South Africa, where the car won dozens of races until 1975. For 1973, Emerson and **Peterson** won a total of 7 races, but were outperformed by **Stewart** who took the title. Emerson left, but Peterson still managed the eventual pole and took three wins with the 72 in 1974, as the new JPS 9 proved to be a failure. For 1975 Lotus continued to race the 72, by then in E configuration, but the car was no longer competitive. The 72 also won a few non-championship F1 events in Emerson's and **Ickx**'s hands.

The McLaren M23 (Alejandro de Brito)

Emerson's move to **McLaren** proved to be shrewd, as the M23 would be competitive for quite a while. Introduced in 1973, in South Africa, the M23 scored a pole right away, which turned out to be **Hulme**'s single pole ever. The car won 3 races in 1973, driven by Hulme and Revson, and as the team lured **Marlboro** sponsorship away from **BRM** and

Emerson from Lotus, the future looked bright as McLaren won the first two races of the 1974 season, plus a non-championship race in **Brasilia**. It did not turn out as easy as it looked, for **Ferrari, Brabham** and **Tyrrell** proved to be worthy competition, but in the end Emerson won the title, his second and McLaren's first. The Brazilian remained in the team in 1975, partnered by **Jochen Mass**, and the trio won two races, even though the car appeared inferior to the new Ferrari 312T for most of the season. Emerson left at the end of the year, but McLaren signed **James Hunt,** who showed the car's late season form was no fluke. After battling royally with **Niki Lauda** in a season of ups and downs, Hunt won the 1976 title including six races. Early attempts to use the M26 in 1976 (driven by Mass) were disappointing, so the works continued to use the M23 in the early 1977 season races. Hunt did not win any races but scored three straight poles in 1977 using the M23. The M23's last competitive outing in the World Championship was second in Sweden, 1977, driven by Mass. M23s were sold to privateers, who occasionally used it in World Championship events. **David Charlton** raced one in his home event a couple of times, while BS Fabrications ran M23s during the 1977 and 1978 seasons, including in the driving strength **Nelson Piquet**. Melchester Racing used an M23 to win the 1978 British Group 8 championship, and entered the car in the British Grand Prix that year, without success, while Spaniard **Emilio de Villota** bought the one-off McLaren M25 (Formula 5000) and converted it to M23 F-1 specification, running the car in several World Championship events in 1977 and 1978, under the **Iberia Airlines** and **Centro Asegurador** F1 banner, without success. The car did win a few races in the minor British series, though.

The **BRM P160** was around for four seasons, 1971 to 1974, and was BRM's last breath of competitive flair. The car designed by **Tony Southgate** won a couple of World Championship races in 1971, as BRM took the runner-up spot in the Constructor's Cup. For 1972, an overly ambitious set-up meant BRM, which built its own engines and transmissions, entered as many as 5 cars in some World Championship events, under seemingly ubiquitous **Marlboro** sponsorship, as the company sponsored many drivers, races, publications. The scheme included a main team of 3 cars, plus Marlboro-country sub-teams. This meant that BRM used as many as 3 different types in individual races (P160, the old P153 and the new P180), and a great number of drivers which certainly did not boost morale. The high point of the season was **Beltoise**'s win in Monaco, BRM's last. For 1973 BRM downsized its attack to three cars, driven by **Regazzoni, Beltoise** and **Lauda**. The P160 still did reasonably well, as all three drivers led races during the season, while Regazzoni achieved pole in **Buenos Aires**. That was not enough to retain Marlboro and most drivers, and that was the beginning of the end for BRM. The P160 continued to be used in 1974, as the new P201 was not available, and Beltoise actually scored a couple of points in a P160 car in Argentina. **Pescarolo** and **Migault** drove P160Es for most of the season, without success.

A few Lotus F-1 cars were sold in South Africa and won dozens of races. Here is Charlton. (Russell Whitworth)

NATIONALITIES

Drivers from dozens of countries took part in the World Championship from 1950 on, including all habitable continents. Drivers from countries which held races tended to dominate the entry lists, especially in the early history of the championship, but that has not been necessarily true later. The list below indicates (*) drivers from the country won a race, (!) driver from the country won a championship.

EUROPE
Great Britain
    England * !
    Scotland * !
    Wales
    Northern Ireland *
Ireland
France * !
Italy * !
Austria * !
Germany * !
Finland * !
Sweden *
Holland * !
Belgium *
Spain * !
Monaco *
Portugal
Switzerland *
Poland *
Czech Republic
Hungary
Liechtenstein
Denmark
Russia

NORTH AMERICA
USA * !
Canada * !

CENTRAL AMERICA
Mexico *

Eliseo Salazar, here driving a March, was the only Chilean driver to make it to Formula 1 (Kurt Oblinger)

SOUTH AMERICA
Argentina * !
Brazil * !
Venezuela *
Colombia *
Uruguay
Chile

AFRICA
South Africa *!
Morocco
Rhodesia (now Zimbabwe)

ASIA
Japan
Malaysia
China
India
Thailand (also referred as Siam)
Indonesia

OCEANIA
Australia *!
New Zealand *!

A few countries that have held World Championship events have never been represented by drivers in the World Championship. That specifically applies to Middle Eastern countries, which seem unable to bring up suitable talent up the ladder, even though money is not a hindrance.

These countries are:

Turkey
Singapore
South Korea (even though some insist to say **Jack Aitken** is Korean)
Azerbaijan
San Marino (races held in Italy)
Luxembourg (race held in Germany) (**Bertrand Gachot** was born in Luxembourg, but raced under the French flag)
Abu Dhabi
Bahrain
Saudi Arabia
Qatar

The issue of driver nationalities can often lead to great many discussions, as drivers may have been born in one country, but hold citizenship in another. **Keke Rosberg**, for instance, was born in Sweden, but has Finnish citizenship, because his parents were studying in Sweden when he was born. **Jochen Rindt** was born in German territory to

Austrian parents, while **Nelson Piquet Junior** was born in Germany, but raced under Brazilian citizenship. **Pietro Fittipaldi,** born in the United States, also races under Brazilian flag.

Many drivers have double citizenship, such as **Hernando da Silva Ramos**, who had both Brazilian and French citizenships. Not surprisingly, Brazilian sources call him Brazilian, French sources, well, French. Everybody else has to pick sides.

Some people have gone as far as saying that **Mario Andretti** is a Croatian! There is some element of truth to that, because the village where Mario was born in 1940 was Italian territory during World War II but became part of Yugoslavia (thus Croatia) after the war. The fact is that at the time it was Italian territory, and Mario raced under the American flag anyhow.

Harry Schell held both French and American citizenships, but raced under the American flag. Yet, sometimes his citizenship appears as French in some results.

Roberto Guerrero, the first Colombian to race in F1 has acquired the American citizenship since leaving the sport, but ran under Colombian colors.

Hans Joachim Stuck, was known as a German driver in his heyday, but now appears to be using the Austrian citizenship.

A few lesser drivers have also been born in more exotic settings. **Mike Beuttler**, a Briton through and through, was actually born in Egypt. Some biographies insist **Jo Schlesser** was born in Madagascar, while others state he was born in Liouville, Northern France. While it is true that Jo lived many years in Africa, it does not appear as if Schlesser was born in Madagascar after all.

Alex Albon, on the other hand, was born in London, but races under Thai citizenship, as exotic a nationality as they come.

Another nationality appeared ready to be added to the list, Estonia. However, driver **Jüri Vips**, who was in the **Red Bull** roster, was terminated having used racial slurs on a video that found its way on the internet.

A Chinese driver, **Zhou Ghanyu**, has recently joined Formula One. A more detail oriented viewer of races will notice that drivers are identified with abbreviations of their last names on screen stats, only **Zhou** is apparently identified by his apparent first name, **Zho**. Appearances are deceiving. It so happens that in Chinese the family name comes first, so the driver's family name is actually Zhou, rather than Ghanyu.

A driver who raced in the South African race in the 1960s, **Mike Harris** came from Northern Rhodesia, which is now Zambia. So while some statistics lump them into Rhodesia (which is now Zimbabwe) others insist on referring to Zambian F1 drivers.

ABUNDANCE OF DRIVERS OF CERTAIN NATIONALITIES

The abundance of drivers of a given nationality has changed with time, due to a number of reasons: number of teams with the same nationality, economic and political situation of the country, language spoken in the country, sponsor's inclination to spend, nationality of current champions, etc.

In the early seasons of the championship, there was an abundance of Italian drivers in F1, for there were many Italian teams running Alfa Romeo, Ferrari, Maserati and Osca cars. Argentines, which generally could communicate well in Italian, also found many opportunities in such teams. It should be noted that drivers from that time were born in the early part of the century, and most did not speak a second language.   This situation changed by the late 1950s, when British and English speaking drivers prevailed, in fact, from 1958 to 1969 all champion drivers were English speakers. This became less of a factor with time, as younger generations of all nationalities were able to communicate well in English and the language became the lingua franca of Formula 1 and international racing in general. Nowadays anyone who wants to become an international professional race car driver must know English.

By the late 1980s-early 1990s, a number of new Italian teams appeared in F1, such as **Scuderia Italia, Minardi, Coloni, Andrea Moda, Forti**, and even the British Toleman team was bought by Italian owner **Benetton,** plus a large number of Italian sponsors joined the sport. **Ferrari** and **Osella** were also around. In 1989 there were as many as eleven Italian drivers on some F1 grids, as some also raced in British and French teams such as **Williams, Tyrrell** and **Larrousse**. By comparison nowadays Italian drivers are very hard to come by in F1.

In the late 70s and early 80s the most abundant nationality was French, for a different reason: the success of the French racing lower formulae system, largely supported by Elf and **Renault**. At one point as many as seven French drivers raced in F1 at the same time.

When a driver of a given nationality wins the championship, opportunities seem to open to fellow countrymen. This specifically happened to Austrians and Brazilians in the 70s, Finnish drivers after the 80s, and to an extent, Germans after the 90s. Sponsors are more likely to land a hand, plus there are imponderable psychological reasons, a belief that a golden generation of drivers from a given country has emerged, not always corroborated by on track performance.

DRIVING FROM THE BACK OF THE FIELD

Wins from the back of the field are stuff of legend, such as **Stewart**'s win in South Africa, 1973, and **Watson**'s drives in Long Beach (1982) and Detroit (1983) even farther back, in addition to **Sergio Perez'** victory in Sahkir, (2020). In most of these cases, the drivers behind the winner generally started towards the front, more than likely the top 10 qualifiers. That is especially so now that cars rarely break, and the number of finishers is usually very high. However, during the history of the world championship in some races the top 6 drivers at the end mostly started behind the top end of the grid.

Take for instance, the 1975 German Grand Prix. Winner **Reutemann** started unusually far back, in 10th place, and he was followed by the 13th starter, **Laffite**. Pole-sitter **Lauda** came in third, but he was followed by the 16th, 20th and 21st place starters!

In South Africa, 1978, winner **Peterson** also started unusually low, in 12th place. The other top 6 finishers started 11th, 10th, 18th, 13th and 14th.

In Belgium, 1987, none of the drivers that started in the top five places scored. Winner **Prost** started 6th, and in the race was followed by the 10th, 13th, 11th, 15th and 16th qualifiers. The same **Prost** was the only top dog to finish in the top 6 at **Phoenix**, 1989, winning from a 2nd place start. The other points finishers started 14th, 17th, 26th, 25th and 16th.

The pole winner sometimes had better luck than his colleagues at the top of the pecking order. Winner **Gerhard Berger** won from pole in Germany, 1994, but was followed by the 12th, 14th, 17th, 16th and 22nd qualifiers. Australia 1995 was a similar case. **Damon Hill** won from pole, followed by the 12th, 13th, 10th, 14th and 17th fastest qualifiers. It is noteworthy that the second placed driver in both cases was **Olivier Panis**, driving a **Ligier.**

In the 50s the results occasionally sprung some surprises. **Fangio** won the British Grand Prix of 1956 starting his **Lancia-Ferrari** 2nd. He was followed by the 12th, 13th, 21st, 14th and 19th fastest qualifiers.

In conclusion, maybe if the likes of **Caterham, HRT** and **Virgin** had been around earlier, when cars broke in profusion, they might have scored points here and there.

DRIVERS THAT SHOULD HAVE WON

On top of everybody's lists, I suppose, is **Chris Amon**. The New Zealander, who raced from 1963 to 1976 for a large number of teams, including his own, was a top driver from 1967 to 1972, in which time he raced for **Ferrari, Matra** and **March**. He is widely acknowledged as the unluckiest Grand Prix driver ever, missing out on ultimate success quite a few times. A great qualifier, fast in all conditions, Chris dropped out from the Canadian GP of 1968 when leading by one minute. He also led several other times in 1968 but came away with only 10 points at the end of the season. After leading in Spain, 1969, Chris got disheartened with Ferrari and left the team. He joined March in 1970, and although he finished in the points often, coming close to winning both the Belgian and French races, and starting on the front row early in the season, he came home empty again. At Matra it was more of the same.

Chris Amon came close to winning the elusive first GP with Matra-Simca, but came up short. (Russell Whitworth)

The pole positions in **Monza**, 1971, and **Clermont-Ferrand**, 1972, did not result in wins, just praises. In total, Amon led 183 laps in his heyday, more than a large number of Grand Prix winners and even champions. Late in 1973 Amon was hired to drive the third Tyrrell in the Canadian and US races, but **Cevert**'s death meant the New Zealander did not have a chance to convince **Ken Tyrrell** he was a contender for the 1974 ride.

This writer has been mostly impressed with **Nico Hülkemberg's** performances in Formula 1, although I admit that making the case that a driver who never even got a podium deserved to win a F-1 race is a bit hard. I would say, looking at the steady and forceful way he conducted himself driving for midfielders he would qualify as a great win candidate. With a career that stretches back to 2010, Nico is still remembered to replace drivers when the going gets tough. He remains the driver with the more starts without even a podium, a dubious honor at best, but he also led a total of 43 laps, not bad for a non-winner.

Towards the end of his career, Jarier drove for Osella (Kurt Oblinger)

Jean-Pierre Jarier demonstrated great speed and skill in the early South American races of 1975, scoring pole in both Argentina and Brazil, and leading handsomely in the latter race. He did not keep that form during the rest of the season but looked formidable again when the replaced the deceased **Ronnie Peterson** at **Lotus,** late 1978. He led most of the Canadian Grand Prix, but the car broke. This performance brought him a few more years racing for the likes of **Tyrrell, Ligier** and **Osella**, but it simply was not to be.

Another fast Frenchman, **Jean Behra**, raced in the 50s mostly for **Gordini** and **Maserati**, but he also competed in **BRM** and **Ferrari.** He lost the 1957 British GP holding at bay **Fangio, Hawthorn, Moss** and **Collins**, but the clutch failed. In total, he led 107 laps in his Grand Prix career, the 10th best among F1 drivers in that decade, although some Indy-only racers had better numbers.

Jean Behra ended his topline F1 career punching a Ferrari manager

He did win occasional non-championship races, and it was felt that a move to **Ferrari** in 1959 would bring the equipment and environment he needed to win. It all came to naught, for Behra ended up in a fist fight with team manager **Tavoni,** after his Ferrari failed in Jean's home race at Reims, and the Frenchman was obviously fired. That essentially meant the end of his topflight Grand Prix career, but it did not matter. Jean died on a support race in the German Grand Prix weekend soon afterwards.

Andrea de Cesaris was mostly known for being a tad sloppy in his early years, but the fact is that the Italian sporadically led races in fine style, and became quite a dependable racer in the latter part of his career, driving for teams such as **Jordan, Rial, Sauber** and **Dallara**.

While most people remember **Stefan Bellof's** great drive in Monaco, 1984, I am most impressed with **Martin Brundle's** drive in **Detroit** that same year. In the end, the efforts of both drivers were wasted, as **Tyrrell**'s results were entirely excluded, a case of heavy handed punishment from the powers that be in racing. That was Brundle's debut season, mind you. He did eventually have a long Grand Prix career that lasted until 1996, and drove for **Benetton, McLaren, Ligier, Brabham,** and **Zakspeed**, in addition to Tyrrell. He did amass a few podiums and a second place and kept **Schumacher** on his toes in the 1992 season, scoring more often than not.

Stefan Johansson had his chances, driving for two of the top teams in the business at the time, **Ferrari** and **McLaren**. He did quite a professional job driving for both, but qualifying was his weak point during those years, so he always had to come from behind. It should also be mentioned that in this period most winning was done by **Prost, Senna, Piquet, Mansell and Berger**, and it was very difficult for new winners to break through

Two drivers who ended up never winning a GP, came very close to that elusive win in the closest GP finish ever, the 1971 Italian Grand Prix. In fact, the top seven finishers in that race had never won a Grand Prix at that time! In the end, lucky **Peter Gethin** did get his single victory, to the expense of both **Mike Hailwood** and **Howden Ganley**, 4th and 5th on the road. Less than a second separated the top 5, and any of them could have come out the winner. Ganley's car was similar to Gethin's, a 12-cylinder BRM P160, while the other drivers in the mix, **Peterson, Cevert** and Hailwood, drove **Cosworths**. Ganley would never have a chance again, but Hailwood would still race towards the front on a couple of occasions. Cevert would win his only GP two races ahead and Peterson would win quite a few starting in 1973.

Peter Gethin won the 1971 Italian Grand Prix leading three laps (Rob Neuzel)

Hailwood, a bike champion, had raced in F1 in the early 60s and returned to F1 in that race. In the first three GPs of his return, Mike the Bike led a race, dropped from 2nd in the 1972 South African Grand Prix after starting fourth and scored fastest lap in that race. He would go on to finish 2nd at Monza, 1972, being quite close to winner **Fittipaldi**. He was the best Surtees performer in the history of the marque, but a poor 1973 season did hurt his reputation somewhat. He had a last chance at stardom driving a third Yardley sponsored **McLaren**, but a mid-season accident ended his Grand Prix career.

Other good performers who never got wins were **Eugenio Castellotti** and **Jackie Oliver**. Castellotti finished 3rd in the 1955 season, and would usually be very aggressive in the initial phases of races, with great pace and flamboyance. Oliver led his first race in 1968 and his last one in 1973, in total leading 36 laps driving for Lotus, BRM and Shadow. Some say he was the winner of the 1973 Canadian Grand Prix, mired in timekeeping confusion only made worse with the first-time deployment of a pace car.

Castellotti, here driving a Lancia D50, did not live long to win a Grand Prix, although he had talent and speed (Alejandro de Brito)

Some drivers simply did not have enough time to score that elusive GP win. I would include in this small group **Stefan Bellof, Tony Brise** and **Ricardo Rodriguez**, all of whom had untimely deaths and immense talents.

EPONYMOUS TEAMS

With seemingly infinite words to name Grand Prix teams, there were instances in which similar names were given to two or more Grand Prix teams. Here is the story.

THE ASTONS

Before you get into a heated argument with another enthusiast, you should know that the **Aston-Butterworth** car that appeared in the 1952 Grand Prix season bears no relationship at all with Aston Martin. A Formula 2 car as all others in the game that year, this Aston was the creature of **Bill Aston** and **Archie Butterworth**. Based on a **Cooper** chassis, cars were run for both Aston and **Montgomery-Charrington**, and never finished a race, always plagued by fuel system issues. If you are wondering, Butterworth produced the engine.

The 1959 Aston Martin was outdated, by the time it debuted and it was not successful

As for **Aston Martin**, by the time it entered F-1 racing in 1959 it had been a very traditional and competitive racing car builder since the 1910s. Yet, despite having competent drivers **Roy Salvadori** and **Carrol Shelby** on the driving strength, the front engined Aston Martin proved mostly uncompetitive and outdated and the effort did not last beyond the 1960 season. Salvadori did put an Aston Martin on the front row at **Silverstone,** 1959, but got no points.

The current Aston Martin team initially saw life as **Jordan,** back in 1991. The team changed owners and names several times, until being bought by Canadian billionaire **Lawrence Stroll**, and renamed **Racing Point**. After a very competitive 2020 season, which netted a win, Stroll decided to rename the team Aston Martin, whose control he had recently bought. Despite having former 4-time world champion **Sebastian Vettel** on board for the 2021 and 2022 seasons, the revived Aston Martin has failed to live up to expectation, despite the beautiful green color scheme.

THE ATS's

Call it coincidence…One clueless **ATS** team was a lot for GP racing, two a bit too much. The first one was the result of a **Scuderia Ferrari** rebellion, when key personnel including manager **Carlo Chiti**, of latter **Autodelta** fame, and drivers **Phil Hill** and **Giancarlo Baghetti** fled to a new team in 1963, tired of Ferrari politics. The effort called Automobili Turismo e Sport bankrolled by enthusiast **Count Volpi di Misurata**, built a proprietary engine and entered five events in the 1963 season. Both Hill and Baghetti, former Grand Prix winners, failed to make the tatty looking **A.T.S.** car work, and at the end of the year the team gave up. **Alf Francis, Stirling Moss'** former mechanic, took the car under his wing in 1964, changed it a bit and renamed it **Derrington-Francis-ATS**. The car made a foolish appearance driven by Portuguese driver **Mario Cabral** in Italy, where it failed again.

The 70s-80s ATS was created by **Gunther Schmid**, a German industrialist who built **ATS car wheels**, a company that had already sponsored cars in Formula 2 and Super Vee. The team appeared in 1977, initially running the old **Penske** chassis and scoring 6th on debut, and had a number of drivers, designers and managers until folding in 1984. To its credit, ATS occasionally scored points, and sprung some qualifying surprises, especially the **BMW** turbo engined version driven by **Manfred Winkelhock** in its final seasons. Schmid, for one, had a disturbing habit of meddling in the team's operations, which caused the constant change of personnel. Not one to learn a lesson, Schmid, who had sold ATS and bought wheels manufacturer **Rial** in 1987, decided time was nigh to come back to F-1. In a very crowded environment, Rial lasted a short two seasons, but both **De Cesaris** and **Danner** scored miraculous 4th places with the car.

The German ATS team amazingly managed to stay in F-1 from 1977 to 1984. (Alejandro de Brito)

THE LOTUS

The venerable **Lotus** name appeared in Formula 1 for the first time in Formula One in 1958, on a car driven by **Graham Hill. Colin Chapman**'s company already produced road going cars, which he wisely kept separate from the racing operations. In the 60s and 70s, Colin's obvious genius produced some great innovations in F1 design, as a result of which Lotus became the winningest constructor of the period, winning titles in 1963, 65, 68, 70, 72 and 78. Chapman died of a heart attack in 1982, and the racing team went into a decline, except for a short period when **Ayrton Senna** drove for the team in 1985 to 1987, winning six races. After many flawed designs, change of management, sponsorship and engine partners, the acrimonious end came in 1994, as **Mika Salo** brought his Lotus-**Mugen** into the pits on the 49th lap of the Australian Grand Prix.

The real Chapman Lotus won dozens of races and many championships and raced from 1958 to 1994. The two latter-day Lotus had mixed fortunes and did not survive long. (Alejandro de Brito)

There was always talk of a comeback, and it did take place in the form of **Team Lotus**, a Malaysian team based in England that entered Grand Prix racing in the 2010 season. The team run by **Tony Fernandes** had bought a license to use the Lotus name and logo through **Proton**, a Malaysian carmaker, and went about the business of racing in the back of the grid as did the two other newcomers, **Hispania** and **Virgin**.

The trouble started in late 2010, as Proton terminated the licensing agreement, claiming there were flagrant and persistent breaches of the license by the team. Soon after Fernandes acquired Team Lotus Ventures from **David Hunt**, which he bought upon demise of the original Lotus team in 1994. In the meantime, **Group Lotus** (which built the road cars), had bought a stake in the **Renault** Formula 1 team, and argued that David Hunt had no right to license the name to begin with. So, for 2011, we were gifted with not one, but two teams claiming rights to the Lotus name, one using a color scheme (green and yellow) reminiscent of the **Clark** years (Fernandes), the other dressed in JPS black and gold (Renault). After much legal wrangling and accusations galore, the mess was solved during the year, as Fernandes bought **Caterham** (which incidentally, built Lotus Seven copies) and decided to rename his team as such, while Renault's name was changed to Lotus F1 starting in 2012. It should be noted that Fernandes' cars used Renault engines in 2011, a case of sleeping with the enemy. I suppose the outcome was more favorable, for the second revived Lotus, called Lotus F1, managed to add a couple more victories to Lotus' tally, even though the teams are kept distinct in some statistics and does not add to the original team's heritage. The team morphed into Renault in 2016 and the name dropped, at least for now.

During this time, there was much talk of other old team names being restored, such as **Brabham** and even **March**, but this came to naught.

NOT THE SAME SPELLING, BUT

ALFA AND ALPHA

ARE THE SAME THING

These days, we have on the grid two α teams. One is a very traditional name, one of the most successful Grand Prix constructors from the 20s to the 50s, who made a not very successful comeback in the 70s, and now comes back: **Alfa Romeo**. The other **Alpha Tauri**, a fashion brand created by **Red Bull** in 2016. So two Alfas on the grid, at the same time!

I personally liked the name Toro Rosso, which stands for Red Bull in Italian, which respects the Minardi origin, but I guess the company saw little reason using a name that had no branding per se. As Formula 1 seems to be making a nice transition into the tastes of the younger set, getting a clothing brand involved in F-1 is not a bad idea. Whether the idea was there from the onset, I guess we will not know.

Alpha Tauri has also been lucky in that it won a race in its debut year, 2020, while Toro Rosso took a little longer before winning its single GP at **Monza**, 2008, driven by **Sebastian Vettel**.

As for Alfa Romeo, its logo had been shown on the **Ferrari** F1 cars for a while, before **Sauber** was named Alfa Romeo in the 2018 season. That would be enough to give **Enzo Ferrari** a heart attack, had it happened while he was alive, for it should be remembered that Enzo ran Alfa Romeo's Grand Prix operation until the late 1930s (that is how Scuderia Ferrari started), and they did not part company in the best of terms and became rivals.

Enzo would probably take exception to an **Alfa Romeo-Ferrari**, for that is the engine that Alfa Romeo uses, and that is how the Alfa Romeo F1 is referred to.

Alfa Romeo's future looks bleak, **Audi,** a **VW** brand will take over Sauber in a couple of season. Whether the Italian company will strike an agreement with someone else (for instance, Williams) is a big question mark. In the meantime, have fun with the first letter of the Greek alphabet and don't mix them up.

LAST BUT NOT LEAST, HOW ABOUT TWO AMERICAN TEAMS NAMED HAAS?

To begin with, **Haas** is not even a typical American last name and at best there are a few thousand folks named Haas in the country. Yet, there have benn two American teams named Haas in F1. The first was ran by **Carl Haas**, the cigar chomping American **Lola** importer who decided to enter F1 in 1985, after getting bored with ruling US Formula 5000 and Can-Am for many seasons. The cars were equipped first with **Hart** turbo engines, and then a turbo **Cosworth** V6, and despite having two very good drivers, former champion **Alan Jones** and **Patrick Tambay**, the team made little progress in two seasons. Adding insult to injury the sponsor **Beatrice Foods** changed hands, and the new owners were no lovers of racing, so Haas focused on an Indycar partnership with **Paul Newman**, setting up **Newman-Haas** leaving F1 forever.

Many years later, a dude called **Gene Haas**, who had entered cars in NASCAR for many years, decided to give Formula 1 a go! The owner of a company called **Haas Automation**, the cars' fortunes have oscillated from a very good maiden season, to bad 2020 and 2021 to a promising 2022 beginning, but Gene has not lost faith and has bankrolled most of the expensive operation since 2016. Not only the teams are not related at all, neither are Carl and Gene!

NOBILITY

In the Pre-World War II period finding European nobility in top racing entry lists was relatively common, as it was in many sports. During the World Championship era nobility was replaced by people hauling from a vast number of backgrounds, and in fact, no nobleman has been around the F1 paddock as a driver since the **Johnny Dumfries** retirement.

Von Trips in the awesome shark nose Ferrari (Alejandro de Brito)

The most successful one of the lot was German Count **Wolfgang Von Trips**, who came very close to winning the 1961 World Championship. Von Trips was a fast driver in a number of disciplines, and drove mostly **Ferraris** in his Grand Prix career that stretched back to 1956. The German managed a couple of podium positions before the 1961 season, when the better prepared shark nose Ferrari became the class of the field. By **Monza,** Von Trips led the championship with 33 points and two wins, and started on pole in Monza. Unfortunately, it was not to be. A collision with **Jim Clark** on the early stages of the Italian Grand Prix meant death for the Count and loss of the world championship, for teammate **Phil Hill** won the race and the title. It would take quite a few years for a German to become a factor in F1, in fact until the 90s when **Michael Schumacher** came to the fore.

A Thai prince happened to be the first Asian in Formula 1. **Prince Bira** (also known as **B Bira**) had raced in Europe during the 30s, in fact won voiturette races on occasion, and came back for a swansong during the 50s. In total **Birabongse Bhanuban** took part in 19 races driving **Maseratis, Gordinis** and **Connaughts**) from 1950 to 1954, his best placings 4th in the 1950 Swiss Grand Prix and 1954 French Grand Prix.

Prince Bira (unattributed photo)

Emmanuel de Graffenried was a Swiss Baron, born in France, who had already raced in the Pre-War period. De Graffenried won minor races on occasion, but is better known for winning the non-championship British Grand Prix of 1949, driving a **Maserati**. In the World Championship **Toulo**, as he was known in racing circles, did not have as much luck, but had 22 starts between 1950 and 1956. He was good enough to be called to join **Alfa Romeo** for three races in 1951, finishing 5th in his native Swiss Grand Prix. His best placing was to come in 1953, a fourth in Belgium.

Carel Godin de Beaufort was a Dutch Count who was a bit of nuisance driving a **Porsche** sports car in F1 races in the 50s, but eventually evolved into a point scoring driver, with his **Ecurie Maarsbergen** Porsche in the 60s. In his trusty car Carel managed four sixth places, earned driving against professional privateers and works team drivers. His major achievement, however, might have been qualifying in 8th place for the 1962 German Grand Prix at the daunting **Nürburgring**, quite an effort. Unfortunately, that is the same place where the popular Count found his death in practice for the 1964 Formula 1 Grand Prix.

Alfonso Antonio Vicente Eduardo Angel Blas Francisco de Borja Cabeza de Vaca y Leighton is better known as **Marquis de Portago**. The Spanish nobleman was born in England, and as many noblemen of his time, was an accomplished athlete – a jockey, swimmer, steeplechase and bobsleigh practitioner. He then decided to give motor racing a crack, in 1954, and raced briefly in Formula 1, in 1956 and 1957, always driving for **Ferrari**. His most memorable feat in Formula 1 was sharing the second place **Lancia-Ferrari** with **Peter Collins** in the 1956 British Grand Prix. He also earned a 5th place in his last F1 appearance but died in the 1957 Mille Miglia. Racing was not kind to nobility, it must be said.

A Sicilian Prince, **Gaetano Starabba di Giardinelli** also appeared briefly in the 1961 Italian Grand Prix, driving a 1.5 **Maserati** engined Lotus, lasting 19 laps. Starabba did reasonably well in minor races and has lived to tell about it.

There were a few other members of nobility who raced in Formula 1, including two Belgians. Baron **Charles de Tornaco** first raced in Formula 1 in 1952, starting two races. He practiced for the Belgian Grand Prix of 1953, but did not qualify. The Baron died in an accident practicing in Modena, 1953, breaking his neck and skull, what a gruesome end! The other Belgian nobleman was **Alain de Changy**, a Count who tried to qualify a Cooper in the 1959 Monaco Grand Prix, failing to make the race. He raced mostly in sports cars, and actually finished sixth in the 24 Hours of Le Mans of 1958.

Adolf Brudes Von Breslau was part of German nobility, but no information was found as to his actual title. The fact is that the German began racing bikes in 1919, cars in 1928, and raced until 1968, in other words, his career spanned almost 50 years. His times in F1 were far from outstanding, limited to a single appearance in the German Grand Prix of 1952 driving a Veritas. He did, however, win a few hill climbs over the years, was a Borgward works driver and tested for Auto Union in the 30s

John Colum Crichton-Stuart, at the time **Earl of Dumfries**, later **Marquees of Bute**, was known in racing circles as **Johnny Dumfries**, and raced in Formula 1 in the 80s. The Brit was likely to be the best performing nobleman ever, considering his Formula 3 form, but a morale sapping number 2 berth at **Lotus**, in 1986, alongside one **Ayrton Senna**, put things in perspective. In the end, Dumfries scored only 3 points in the whole season, and decided to leave Formula 1 for the commoners. To his credit, Dumfries did win the **24 Hours of Le Mans with Jan Lammers** and **Andy Wallace**, sharing a **Jaguar** in 1988.

Then there was another prince! **Jorge de Bagration** is known as a Spanish driver, but is indeed Georgian **Giorgi Bagration-Mukhraneli**, heir to the country's theoretical crown. A common sight in Spanish and European sports and touring car racing of the 60s and 70s, Bagration also dabbled in Formula 2 occasionally, which led to his first attempt to enter a F-2 **Lola-Ford** with slightly increased engine in the 1968 Spanish Grand Prix. The entry was not accepted by organizers, who deemed the proposition too uncompetitive. Nevertheless, a Prince does not give up easily, and in 1974 De Bagration again attempted to enter a car, then a proper 3.0 **Surtees-Ford** in the Spanish Grand Prix. There are a few versions to the story but to make it short, the entry was supposedly filed twice, and was misplaced. It also appears that the Prince's pockets were not that deep, and the sponsorship for the car never materialized. Regardless of the version, it would be very unlikely the entry would qualify.

QUALIFYING SURPRISES

The current qualifying set-up adopted in 2010, Friday practice plus 3-step made-for-TV qualifying on Saturday, now with the odd qualifying races, has been around for a short while. For the most part, long qualifying sessions were held on Friday and Saturday before races, giving drivers and teams a chance to set-up cars according to track conditions.

Although the old set-up produced more surprises than the current one, **Nico Hülkenberg's** pole in Brazil, 2010, in the new scheme, must come at the top of the list in any super qualifying performances. Just a few years before any driver would do anything to become a **Williams** driver, but by 2010 the team had started a descent into oblivion. This was Nico's debut season in F1, and that far, he had earned a few points, mostly because the top 10 point scoring had been established that season. Nico got pole, but finished 8th in the race, That was not enough to convince Williams they had talent to nurture, so Nico began his musical chair/reserve career in 2011.

Pastor Maldonado did better than Hulk in the 2012 Spanish Grand Prix. The Venezuelan not only scored pole in his Williams, after the exclusion of fast man **Lewis Hamilton**, but led many laps and won the race to the delight of dictator **Hugo Chavez**, who largely paid the bills. To his credit, this was no fluke result, he led **Alonso** and **Räikkönen**, and did not put a wheel wrong. Unfortunately, he never repeated the performance again, and his wild reputation has stuck.

Another **Williams** surprise in the off-years was **Nick Heidfeld**'s pole position driving a Williams-**BMW** in the European Grand Prix of 2005. Heidfeld was a very good performer, perhaps a perfect number 2 driver, who stayed in F1 between 2000 to 2011. It should be mentioned that Formula 1 was mostly not kind to Formula 3000 champions, and such was the case with Heidfeld. **Massa** also had a pole at Williams, late in his career.

Kevin Magnussen's pole for the 2022 Brazilian Grand Prix (now called Sao Paulo Grand Prix) is also legendary. It was not so much because the Dane was ultrafast, rather he recorded the fastest time before wet weather intervened, the session was red flagged, and the quick guys could not beat the Haas driver's dry time when Q3 resumed. On the other end of the grid was the other Hass of **Mick Schumacher**. Kevin's pole gave him an advantage in the sprint race, which he lead for two laps, but he ended up finishing 8th in the sprint, and retiring in the feature before completing a single lap.

Back in Williams' formative years, when it was still called **Iso-Marlboro,** Williams caused a stir when **Arturo Merzario** qualified third in South Africa, 1974, after finishing 3rd in a non-championship race in Brasilia, Brazil. In **Kyalami** he finished 6th. Another unusual top qualifier in that race was **Carlos Pace**, driving the bad **Surtees TS-16**, who was on the front row but finished 11th. By the way that happened to be **Niki Lauda's** first of many poles, so the entire front of the grid looked excitingly unusual.

The year before **Howden Ganley** actually managed to extract a 10th place grid position in his Iso-Marlboro at **Monaco**. Another surprise in that 1973 race was **Chris Amon**, who qualified 12th in the **Tecno.**

The 70s had many surprises. **Vittorio Brambilla's** fast form ended in pole position in Sweden, 1975. Also at **Anderstorp**, but in 1976, **Chris Amon** put the **Ensign** third on the grid. **Mario Andretti** qualified the **Parnelli** 3rd at **Watkins Glen** in 1974, its second event. **Clay Regazzoni** had **BRM**'s last pole position in Argentina, 1973, which was not a fluke, for he led early in the race.

Carlos Reutemann qualified his **Brabham** on pole in his debut in Argentina, 1972. **Hans Joachim Stuck** started on the front row of the 1977 U.S. East Grand Prix, driving a **Brabham-Alfa,** leading the start of the race under deluge conditions, eventually retiring.

Vittorio Brambilla had a nice - but short – run of success in 1975, but didn't get enough points to show it. (Alejandro de Brito)

Jody Scheckter shocked the establishment when he qualified 3rd in an old McLaren M19 in his home race, 1973, then put an M23 on the front row, in France. The driver still had some rough edges, so his outstanding quali did not yield results.

Jean-Pierre Jarier's form in the **Shadow-Ford** DN5 in the South American races was quite striking. Jarier, whose F1 performances since 1971 had been mostly subdued, got pole on both events, but in the races things were differently. He did not even start in Argentina, after problems in the warm-up. In Brazil Jarier ran behind **Reutemann** in the early stages of the race but took the lead in the 5th lap. He looked all set to win, but then the car broke down with fuel system issues in the 33rd lap. The

Frenchman also posted the fastest lap, but did not reproduce this form during the rest of the year, although his teammate **Tom Pryce** had pole in **Silverstone**. Some even maliciously suggested that **Cosworth** had given Shadow a super-Cosworth, an unlikely scenario.

Perhaps the most shocking qualifying performance of the decade was **Kojima's** 10th in the 1976 Japanese Grand Prix. In the race **Masahiro Hasemi** also allegedly scored the fastest lap, to this date the only driver to have achieved fastest laps in all his starts...That would be a good story, especially considering the **Dunlop** tires, except that Hasemi did not score the fastest lap. A few weeks after the race the organizers published that was a timing error, and **Laffite** had scored the best time of the race. Most sources to this day claim Hasemi had fastest lap that day – even I made that mistake. **Kazuyoshi Hoshino** would qualify one of the Kojimas 11th for the 1977 race.

Tyrrell had pole in its debut, in the 1970 Canadian Grand Prix, and went on to be a top-2 GP team while **Stewart** drove. Stewart also put **March** on pole in its debut in South Africa, in the same season, a record unlikely to ever be broken. On the other end of the spectrum, **Ricardo Zunino** qualified a **Brabham-Cosworth** 9th in the U.S. East Grand Prix of 1979, a form he never came close to repeating in his short GP career.

After finishing 3rd in the initial qualifying session, **Emerson Fittipaldi** posted 5th on the grid for the 1976 Brazilian Grand Prix in his new ride, the **Fittipaldi**. Unfortunately, it did not go so well in the race and he never qualified higher than that in every other race he had in the family team, not even when he finished 2nd in the 1978 Brazilian Grand Prix.

There were some surprises in the 80s as well. **Jan Lammers** qualified the **ATS**, a very bad car, in 4th place at **Long Beach**, 1980. **Manfred Winkelhock** also shocked when he ended 5th in the Detroit timesheets, 1982, also driving an ATS. **Teo Fabi** posted pole in the **Toleman-Hart** in the 1985 German Grand Prix. **De Cesaris**, known for his unsteady form in his early career, was extremely consistent in 1988 when he qualified the **Rial-Ford** 12th in four straight races. **Stefan Johannson** was mostly known as a bad qualifier, yet, he put his Ferrari on the front row in the 1985 German Grand Prix. **Ligier** had seen better days by 1985, so **Philippe Streiff**'s 5th place start on the European Grand Prix was somewhat shocking.

Colombian **Roberto Guerrero** seemed to like the Detroit track quite a bit. Driving for **Ensign** in 1982, he posted the 11th fastest time in qualifying. In the next season, driving the **Theodore** (really the new name adopted by Ensign which merged with Theodore), he again had the 11th fastest time in the same race!

During the oversubscribed entry years some small teams came up with surprising qualifying exploits in the 80s and 90s. One such team was **Scuderia Italia**, which ran **Dallara** chassis with **Cosworth** and then **Judd** engines. **Alex Caffi** had a good run in 1989, qualifying third in Hungary, 9th in Monaco and 6th in Detroit. **De Cesaris** also put a **Dallara** in 3rd place on the grid in **Phoenix**, 1990, while **Emanuelle Pirro** qualified 7th in Hungary, 1991.

Olivier Grouillard qualified an **Osella** in 8th place for the 1990 **Phoenix** race, the marques' best quali ever.

Minardi, which had a bad beginning with the **Motori Moderni** engine in its early days, had shown speed on occasion, especially in the hands of **Pierluigi Martini** who raced for the team in many seasons. In the late part of the 1989 season Martini qualified 5th in Portugal, 4th in Spain and 3rd in Australia, then achieved what seemed impossible just a few seasons back: front row in the 1990 **Phoenix** race, a GP that seemed to favor the small Italian teams. **Morbidelli** also put a **Minardi-Ferrari** 7th on the grid in San Marino, 1991, and **Luis Perez-Sala** qualified 9th in Portugal, 1989.

Williams was a top team in 1988, in fact had won the 1987 championship, but had lost the **Honda** turbo contract to **McLaren** for the 1988 season. Just a few teams had turbo engines that year, which were much more powerful than the normally aspirated engines that were supposed to be the norm in 1989. Yet, in the first race of the year **Nigel Mansell** qualified the **Judd** equipped Williams on the front row in Brazil, in the engine's debut.

In the latter part of the 90s surprises were fewer, as the number of teams decreased and a more established order emerged: it simply became more difficult for low budget, small teams, to cause a stir. Yet, **Gabrielle Tarquini** qualified his **Fondmetal** 11th and 12th in the 1992 Belgian and Hungarian races, and **Karl Wendlinger** had the 9th starting position in the **Ilmor** engined **March** in Spain and Brazil, in the same season.

In the 60s, Mexican **Moises Solana** put his **Lotus** 9th on grid for his home race, while another Mexican, a very young **Ricardo Rodriguez**, qualified his shark nose **Ferrari** 2nd for the 1961 Italian Grand Prix. **Johnny Servoz-Gavin** surprised everybody when he qualified his **Matra** with a **Ford** engine on the front row at Monaco, 1968,in 2nd place. **Ronnie Bucknum** qualified his **Honda** 6th for the 1965 Italian Grand Prix, by far his best qualifying effort. **Vic Elford** a sports car specialist qualified his **Cooper-BRM**, a reliable but slow car, 5th at the **Nürburgring**.

The German race welcomed both Formula 1 and Formula 2 cars in the 1966 to 1969 seasons, and **Jacky Ickx** astounded the world when he had the 3rd fastest time in the 1967 German GP, driving a 1.6 Formula 2 **Matra-Ford**. He qualified faster than most F-1 cars, but the F-2s started separately from the F1s, so on the road he started 18th. In the actual race Ickx did not fare well.

I suppose the most surprising qualifying effort of the 50s was **Roy Salvadori's** 2nd in the 1959 British Grand Prix, driving an antiquated front engine **Aston Martin** that did not perform well anywhere else. **Joakin Bonnier's** pole in the Dutch Grand Prix of 1959 surprised pundits, but any suspicions of wrong timekeeping were dispelled as the Swede won his, and **BRM**'s, first race and went on to lead 139 laps in his long career. Incidentally, in the book <u>400 Cavalli nella Schiena</u>, Fangio reckoned in a 1961 interview that Bonnier was best driver in F1, after Moss, quite a praise coming from the great driver.

Giancarlo Fisichella's pole position in Belgium, 2009 in the **Force India** has been discussed elsewhere, but the team continued to surprise as it changed owners. **Lawrence Stroll** bought the team and renamed it **Racing Point** and brought son **Lance Stroll** on board. Having a wealthy dad can be a curse in racing, and there are many that feel Lance Stroll, Lawrence's son, is totally devoid of talent, forgetting that one day he was part of the **Ferrari Academy** in 2010-2015 and won many lower formulae championships in succession, from 2014 to 2016, including European Formula 3. Lance has performed very well on occasion in F1, including a pole position in the 2020 Turkish Grand Prix. Time will tell if he will ever excel, although it looks highly unlikely at this juncture.

...AND SURPRISING FASTEST LAPS

Perhaps one of the most surprising facts about fastest laps is that one of the fastest drivers, if not the fastest, to ever race in F-1 posted a relatively small number of fastest race laps. **Ayrton Senna** had only 19 fastest laps in the years he raced F-1, even though he got 65 poles and 41 wins. I reckon that Ayrton's gift was consistently lapping fast, rather than running the fastest lap of the day. On the other hand, **Kimi Räikkönen** had a real flair for the one-lap wonders and is currently third in the all-time list. Currently, at the time of writing the only actual threat to him is 28 fastest laps short, so he is bound to be out there for quite some time.

On that note, even a driver who had not scored a single point in F1 actually managed to get a fastest lap on the records. That is the case of **Brian Henton**, a British driver who won championships in both F-3 and F-2, but never got a proper ride in F-1, in spite of racing in the category on and off between 1975 to 1982. His glorious day came in in the British Grand Prix of 1982, when Brian raced for **Tyrrell**.

Masahiro Hasemi, on the hand, raced a single time in F1, in the 1976 Japanese Grand Prix, where he drove a **Kojima-Ford**. Most sources claim that Hasemi scored the best time, however, a few weeks after the race the organizers sent a press release indicating **Jacques Laffite** was the fastest man of the day.

Other Japanese drivers fit into this group, such as **Satoru Nakajima**, who drove for **Lotus** in 1989 and got the fastest lap in the Australian Grand Prix, while **Kamui Kobayashi** also managed a single fastest lap in his curriculum, driving a Sauber in China, 2012. Unfortunately he would soon be at **Caterham,** where fastest laps were a definite no-no.

Richard Attwood was a very fast sports car driver, who raced relatively little in F-1, but not only got the fastest lap driving a **Lotus** at **Monaco**, 1968, also finished 2nd.

Another driver who got a fastest lap driving for Lotus in 1968 was **Jackie Oliver**, who also led laps in a few races and finished in the podium during his career.

Bruno Senna's career in F1 was far from stellar, yet, he posted the fastest lap in the 2012 Belgian Grand Prix, driving for **Williams**. Another unlikely candidate from this era was Russian **Vitaly Petrov**, who got the fastest lap in the 2010 Turkish Grand Prix, driving a **Renault** in his first season.

As for other unusual fastest laps in the 2000s, one can note **Pedro de la Rosa's** in the 2005 Bahrain Grand Prix, **Timo Glock's** in the 2009 European Grand Prix and **Adrian Sutil's** in the 2009 Italian Grand Prix. While De La Rosa got his fastest lap driving for top team **McLaren**, Glock and Sutil got theirs driving for teams without great records, **Toyota** and **Force India**.

Nico Rosberg was an unknown quantity when he debuted for **Williams** in Bahrain, 2016. Although he was a GP2 Champion, there were some detractors that felt his last name was the reason for his being near a Formula 1 grid. Nico did an outstanding job by posting fastest lap in his debut race and finished seventh. He went on to score many other fastest laps, wins and poles in his career at Mercedes, shutting up those early detractors big-time.

Henri Pescarolo had the fastest lap in F-1 for a long period, by posting the fastest lap in the 1971 Grand Prix driving a privateer **Williams March**, at 153.49 mph (247.02 kph). Despite being around from 1968 to 1976 Henri never achieve much success in F1, although he got a 3rd place at **Monaco,** 1970. Yet, he did win dozens of sports car races, including **Le Mans** four times, plus Formula 2 races as late as 1973 and raced into the 2000s.

Mauricio Gugelmin's single fastest lap driving a **March** in the 1989 French Grand Prix was also set in unusual circumstances. In the first start, **Gugelmin** had a terrifying crash, but mustered enough courage to start again and be the fastest on track. What a day!

In the 90s **Bertrand Gachot** proved the Jordan was a very good car by getting the fastest lap in Hungary, 1991. As for **Roberto Moreno**, his fastest lap came in Belgium, 1991, which turned out to be his last race for **Benetton**. He would drive for **Jordan** in the next race, taking Gachot's place, who was in jail in England, having been convicted for aggravated assault.

Another **Benetton** driver who put one fastest lap in the books was Austrian **Alexander Wurz**, in the 1998 Argentine Grand Prix. Wurz was a fast driver whose GP career never really took off. He managed to stay around between 1997 to 2009, and even got a last full season driving for **Williams** in 2007, after many seasons as a test driver and a single race start for **McLaren**, between 2001 and 2006.

In the 50s, Argentine **Roberto Mieres** posted the fastest lap in the 1955 Dutch Grand Prix. One year before, no less than three Argentines set the fastest lap in the British Grand Prix, **Fangio, Gonzalez** and **Marimon**. Additionally, four other drivers scored fastest laps that day, **Ascari, Behra, Hawthorn** and **Moss**, and given that fastest lap entitled the driver to a 1 point, this was divided into seven different parts, resulting in some weird looking score at the end of the season: Fangio had 57.14 points, Gonzalez, 26.64, Hawthorn 24.64, Marimon and Moss 4.14 each, Ascari 1.14 and Behra, 0.14. The unlucky Marimon died in the next race, at the **Nürburgring.**

John Surtees was a very successful F1 driver when he posted his last fastest lap, in South Africa, 1970. The unusual aspect of the feat was that he got it driving a self-entered old **McLaren**, which he would soon replace by his own **Surtees** machines. The Surtees marque got some fastest laps during its stint in F1, from 1970 to 1978: **Pace** posted fastest laps two races in a row, at Germany and Austria, 1973, while **Hailwood** got fastest lap in South Africa, 1972.

Carlos Pace in the Surtees TS-14 in 1973 (Rogerio da Luz)

A little bit too late, one might say, **Daniil Kyvat** scored his single fastest lap in his first race for **Toro Rosso**, when he was demoted to make room at **Red Bull** for wunderkind **Max Verstappen** in Spain, 2016. His achievement in the race was barely noted, as the 18-year old simply won his first race driving for Red Bull, turning the Russian's performance very secondary indeed.

Marc Surer can be proud of being the only driver to ever get a fastest lap driving an **Ensign**, at the 1981 Brazilian Grand Prix, even though drivers of the caliber of **Amon, Ickx** and **Regazzoni** had driven for the marque at one type or another.

Clay Regazzoni was one star driver who also raced an Ensign
(Alejandro de Brito)

Italian **Stefano Modena** was known to be temperamental, and always had the looks of a startled person facing a huge problem. His personality traits cut his career short, for he drove in Formula 1 between 1987 and 1992, in total 70 entries. He was, however, an extremely fast driver and on occasion extracted excellent performances from bad to fair cars. In 1988, for example he drove for **Euro Brun**, one of the worst of the wave of new teams to come into the sport, and managed to qualify more often than not, in fact started the Canadian GP in 15th. In two years at **Brabham**, he had eight top 10 starting positions in 1989, and finished 3rd at Monaco. In 1991 he moved to **Tyrrell,** and started the Monaco Grand Prix on the front row, posting the second fastest time with the **Honda** powered machine. He retired from that race, but then finished 2nd in the next race, in Canada. He started in the top 10 in nine races, and his worst qualifying effort was 14th. Despite such obvious talent, only **Jordan** showed interest in him for the 1992 season, but the **Yamaha** engined car was very slow that season, a shadow of the 1991 car, essentially putting the nail on the coffin of Stefano's GP career.

The dramatic Belgian Grand Prix of 2022 was run under a red flag for two laps, before being permanently called-off. As a result of the permanent red flag, no fastest lap was recognized, but the one driver who scored the fastest lap that day was **Nikita Mazepin**, who drove a Haas. He actually recorded it in the 2nd lap, while the race result that stood was the status quo at the end of the first lap. That would turn out to be the Russian's crown achievement in his F1 career, but it was not to be.

Last but not least **Dr. Jonathan Palmer** managed to squeeze a fastest lap out of his Formula 1 career, driving a **Tyrrell** in the 1989 Canadian Grand Prix, his last season in the category. There is a reason for that: it began to rain very hard after the 16th lap of the race, and all fastest laps were recorded before the 15th lap. Palmer just happened to have scored his fastest lap before the deluge, but he was not particularly fast otherwise. Luck also plays a role in the fastest lap game.

UNIQUE PODIUMS

The first goal of a F1 driver is scoring their initial points, even a single one will do. One can say that scoring a point is not as difficult as it used to be, for now the top 10 drivers get points, out of twenty starters. Compare that with points helpings from first to fifth in some races in the fifties with more than thirty starters. That is one way to look at it, except top cars nowadays rarely retire, so even getting to the top ten can be a very tall order. Just ask the three second-tier teams introduced to Formula 1 in 2010...

A driver's second goal is standing on the podium, in other words, finishing in the top 3 in a race. For many of these drivers the podium has been the crowning achievement of sometimes long careers in motorsport. Some had very short careers, though.

Take **Alfonso de Portago**, who finished 2nd sharing a **Ferrari** with **Peter Collins** in the 1956 British GP. In a race of attrition where many top performers fell by the wayside, Alfonso got his name on the books. He started racing in 1954, and by 1957 had died in an accident in the Mille Miglia.

Onofre Marimon also got a single podium from a short career, also at **Silverstone**, in 1954. He was also one of seven drivers to share fastest lap and the point that came with it, in other words, 0.14 point.

Italian **Dorino Serafini** had a single race in F-1 and finished second sharing a **Ferrari** with faster driver **Ascari** in the 1950 Italian Grand Prix.

As for Swiss driver **Rudi Fischer**, he had the pleasure of getting two podiums under his belt driving his private

Ferrari 500 to 2nd in the 1952 Swiss Grand Prix, and 3rd in the German Grand Prix (the one with tons of starters, by the way). He was far from a professional driver, just a fast restaurateur having fun.

Adrian Newey's designs for **March** and **Leyton-House** showed a star designer was emerging in Formula 1. **Ivan Capelli** got a couple of 2nd places, Portugal in 1989, and in France, 1991, while **Mauricio Gugelmin** finished 3rd in Brazil, 1989.

Stefan Johannson got quite a few podiums driving for the likes of **Ferrari** and **McLaren** but makes this list because of his 3rd place driving an **Onyx-Cosworth** in the 1989 Portuguese Grand Prix.

Piers Courage was highly reckoned as a future star, and his two second places driving a privateer **Williams** entered **Brabham** in 1969 are a proof of that. Unfortunately, he would not survive beyond the 1970 season, when he drove the slow **De Tomaso**, dying in **Zandvoort**.

Brian Redman got a single podium in F-1, driving his heavy and slow **Cooper-BRM** to third in Spain, 1968. His teammate **Lucien Bianchi** matched the performance driving his car to 3rd in **Monaco** that same year. These were the last noteworthy **Cooper** performances in F-1 as the marque would race a last time in 1969.

George Follmer's first two F-1 races augured well for the future. In the 1973 South African Grand Prix, in **Shadow**'s debut, the American finished 6th. He followed that up with third in Spain, but did not do well in any other races during the year. Reportedly, he felt ill at ease with the F-1 ambience. His teammate **Jackie Oliver** also surprised with a podium in Canada, after having a most obscure season.

Shadow impressed at times in its first season, 1973 (Alejandro de Brito)

Another American, **Michael Andretti**, came to F-1 with a wonderful resume. Slated to drive the second **McLaren**, as a teammate to **Ayrton Senna**, Michael also seemed out of place in the category and did poorly almost the entire season. His last race for the team, at **Monza**, 1993, yielded a podium from 3rd. But that was too late.

Emerson Fittipaldi had several podiums in most of his seasons in Formula 1. However his last podium was significant, for Emerson started last in **Long Beach**, and finished 3rd, in 1980. A couple of races earlier **Keke Rosberg** also got his first podium in F-1, driving his **Fittipaldi** (née **Wolf**) to third in Argentina.
Let us go back to the 50s. **Cesare Perdisa** shared a works Maserati with **Stirling Moss** in the 1956 Belgian Grand Prix and got third place from his troubles. The two drivers shared cars in more than one occasion during that season, but **Fangio** still prevailed.

Ron Flockhart started 23rd in the 1956 Italian Grand Prix, but made his way to the front of the field, finishing 3rd. All the more remarkable, he drove a **Connaught**, a marque that was enjoying its last days as a works team.

Fittipaldi at Long Beach, 1980 (Kurt Oblinger)

Many Argentines besides mega champion **Fangio** showed great pace in F-1 during the 50s, one of them being **Carlos Menditeguy**, who finished 3rd in the 1957 Argentine Grand Prix. Carlos' last start in the category was in 1960.

Harry Schell managed to stay around the F-1 circus for quite a while, but got his best result in his 9th season, a second place in the 1959 Dutch Grand Prix. He would still race in 1959 and 1960, so he competed in 11 seasons altogether. More about him in long career spans.

Rubens Barrichello was considered a great promise when he signed up for the **Stewart Grand Prix** team for the 1997 season, but the promise would be fulfilled in earnest in his **Ferrari** and **Brawn** years. One great performance from Rubens that season was the 2nd under the pouring rain at **Monaco**, which turned out to be the team's and his only great result in that season.

Another great promise was **Tim Schenken**. The Australian was considered World Champion material around 1968, in fact won in all categories he tried in Europe, except

Formula 1. His best result was a 3rd earned in the 1971 Austrian Grand Prix, driving a **Brabham**.

As for promises, **J.J. Lehto** looked like the real article in his early days and was very disappointing as a second driver to **Michael Schumacher** in 1994. Notwithstanding, the Finn got a 3rd place driving a **Dallara** in the 1991 San Marino Grand Prix and would enjoy success as a sports car driver.

Some other noteworthy podiums were **Morbidelli's** third driving a **Footwork Hart** in the Australian GP of 1995, **Eric Bernard's** third in a **Ligier**, in the 1994 German Grand Prix, **Philippe Streiff's** third in Australia, 1985, also driving a **Ligier**, and **Mark Blundell's** third driving a **Tyrrell-Yamaha** in the 1994 Spanish Grand Prix.

James Hunt's brilliant podiums driving a privateer entered March in 1973 also deserve mention. Hunt finished 3rd in Holland and 2nd in the USA, in the last case very close to winner **Ronnie Peterson**. It should be noted that many drivers ran in March cars that year, and Hunt, a rookie, was the only one that make the car go fast, in a privateer car for a debuting team.
(Rob Neuzel)

Talking about **Yamaha** engines, one cannot fail to mention **Damon Hill**'s outstanding drive in the 1997 Hungarian Grand Prix, where he led most laps in an unimpressive **Arrows**, then finished second. Those that belittled Damon's skills during his **Williams** years must have seen the light on that day.

Eddie Cheever two podiums in the 1982 season, driving a **Talbot (Matra)** engine **Ligier** are also noteworthy, considering it was a most competitive season, where 11 drivers and 7 constructors won races. This he achieved in two U.S. races, at **Detroit** (2nd) and **Las Vegas** (3rd.).

Another driver who did well on home soil was **Aguri Suzuki**, who finished 3rd in the 1990 Japanese Grand Prix driving a **Lola-Lamborghini** for the **Larrousse** team, having started 10th. Aguri would later become a constructor in his own right, albeit not a successful one.

The 2022 British Grand Prix podium drivers, Perez, Sainz and Hamilton (Red Bull)

FERRARI HITS AND MISSES

As the only team that has been in Formula 1 since 1950, **Ferrari** is bound to have more stories than all others. Add to that the drama and politics that involved the Italian team, which is worshipped by a very emotional fanbase, and you have the Ferrari hits and misses. Although Italian drivers formed the core strength at Ferrari in the early 50s, by the end of the decade the team's fortunes were in the hands of non-Italian drivers. To an average Italian, Ferrari's performance is more important than Italian drivers, so many care less about the lack of Italian drivers in the Scuderia to begin with. Notwithstanding, Ferrari has given chances to Italian drivers in the last few decades, which represent most hits and misses in this list even though it does feature some non-Italians.

Since 1950 there have been some excellent Ferraris, and some bad ones. This is a bad one, the 1973 B3. (Paul Kooyman)

Andrea de Adamich was an **Alfa Romeo** driver who spent a couple of seasons driving for **Ferrari.** In the World Championship, Andrea surprised many by outqualifying both **Amon** and **Ickx** in the 1968 South African Grand Prix, at 7th, although an accident finished his race. However, in the Race of Champions de Adamich had an accident that sidelined him for most of the season, and the Italian never again had a chance in one of the Prancing Horse's cars in F1. **MISS**.

Nanni Galli was also an **Alfa Romeo** driver, who was given a chance by **Ferrari,** as a substitute for **Clay Regazzoni** in France in 1972, for the Swiss had injured himself in a soccer game. The Italian, who was racing for **Tecno** in F1 that year, did not do well. He qualified a lowly 19th and finished 13th and was never given another chance. So he went back to **Tecno. MISS**.

In the next race, the British GP, **Regazzoni** continued his road to recovery, and **Ferrari** called upon the services of **Arturo Merzario**, who had won a couple of World Sports cars races for Ferrari that season. Little Art qualified 9th in his GP debut, and finished 6th, enough to convince the Commendatore to give the Italian a race seat for 1973. **HIT**

The 1982 F1 season was the most dramatic ever, and a lot of the drama involved **Ferrari**, as you will see elsewhere in the book. Early in the season, **Gilles Villeneuve** died at **Zolder,** after feeling short-changed by the team in the poorly supported San Marino race. So Ferrari was down to one driver, and out of respect for the deceased Canadian, ran a single car for a while. As it became clear **Didier Pironi** was the favorite to win the championship, a second driver was hired to support the team leader. This was **Patrick Tambay**, who disappointed at McLaren in 1979. Tambay did not star in his first races, but when called upon he came up with the goods. In Germany it was time for Pironi to get badly injured in practice, however Tambay did win the race, and protected his countryman's position in the championship, to no avail, as it turned out. **HIT**

Mario Andretti had been a **Ferrari** GP driver in 1971 and 1972, and did very well, winning his first GP in South Africa and then the Questor F1/F5000 race in the USA. By 1982 he had already been world champion, won quite a number of races for **Lotus** and seemed to be nearing the end of his GP career. In the dramatic 1982 season Andretti was called to substitute the injured **Pironi** at Ferrari in **Monza**, and surprised many by scoring pole. In this next to last race he finished third, getting another crack in **Las Vegas** where he did not do so well. **HIT**

Ignazio Giunti was another **Alfa Romeo** factory driver who was given a GP chance by **Ferrari**. That came in the 1970 Belgian race, and he finished a very good fourth. Things would not work out for Ignazio in the long-term. He got three more race starts in the 1970 season, but in the 1000 km of **Buenos Aires** sports car race of 1971 he had an accident, hitting **Beltoise's Matra**, and died on the spot, finishing a promising career. **HIT**

Mika Salo was **Mika Häkkinen's** main challenger in their Formula 3 days, in 1989, but was always outperformed by his countryman. Eventually Salo was given F1 chances, but most had been in midfield teams, which gave him little possibility to shine. A good chance emerged in 1999, as **Michael Schumacher** got injured mid-season. Ferrari was not about to repeat the 1982 mistake of running a single driver in adverse conditions, so Salo was called in, and provided support to **Eddie Irvine's** title bid. The Finn almost won the German Grand Prix, but let the Irish driver pass him. **HIT**

Being a test and reserve driver provides a theoretical world of opportunities to drivers, but has been mostly frustrating for a great many drivers. **Luca Badoer** looked promising in the lower formulae, but his chances in F1 in the 90s were in slower teams. So he became a **Ferrari** test driver, a task he took with gusto for many years, waiting for a chance. That chance came in 2009, as **Felipe Massa** got injured in qualifying for the Hungarian Grand Prix. Badoer was then called to replace the Brazilian in the European and Belgian Grand Prix, with poor results. **MISS**

Giancarlo Fisichella had some success in Formula 1 driving for **Jordan** and **Renault**, but in 2009 he was at best a midfield runner in the **Force-India** team. In Spa, Giancarlo surprised everybody scoring a freak pole in the low downforce car and finishing second. *Fisico* was called to replace **Luca Badoer** at **Ferrari**, who in turn had substituted for the injured Massa, as of the Italian Grand Prix. The Italian's performance in Ferrari was woefully disappointing, failing to impress and to score, and he would never get another F1 chance at Ferrari or elsewhere. **MISS**.

Nicola Larini had proved to be a fast driver in small Formula 1 teams and was eventually retained by **Ferrari** as a reserve and test driver. A big chance appeared in the early part of the 1994 season, as **Jean Alesi** was sidelined with injury. Larini drove for the Prancing Horse in the Pacific and the San Marino Grand Prix, qualifying well for both races. In the fateful **Imola** race Larini finished an impressive second, but had to give the seat back to Alesi in the next race and was never again used by Ferrari. **HIT**

Gianni Morbidelli stayed around F1 circles for a while, but as many Italians in his generation, was never given a top ride on a consistent basis. The Italian did the first 15 races of the 1991 season at **Minardi**, often outperforming the car, which for the first time had Ferrari engines. As **Alain Prost** fell out of favor at **Ferrari,** after referring to the cars as trucks, a berth appeared in the last race in Australia. As it was, the race was a very short rainy affair, but Morbidelli did finish 6th and got half a point for his trouble. He was never used by Ferrari again. **HIT**

Lorenzo Bandini was obviously **Ferrari's** darling in the mid-60s, and was performing really well in the first year of the 3.0 Formula. His countryman **Scarfiotti** was seen more as a sports car and hill climbing specialist and was given few opportunities in Grand Prix racing. In fact, Ferrari saw it fit to give the driver a 2.4-liter car in Germany, 1966, but upgraded his ride to a full 3.0 liter car for **Monza**. Ludovico took advantage of this chance and won the race, outperforming teammates **Michael Parkes** and **Lorenzo Bandini**. **HIT**

Ivan Capelli – see the next section on Major Disappointments. **MAJOR MISS**

MAJOR DISAPPOINTMENTS

David Walker, 1972

David Walker won dozens of Formula 3 races in 1970-71, driving a **Lotus** works car, so one could assume that such a dominating performer would morph well into a Grand Prix driver. The Australian had a few F1 outings in 1971, including a drive in the interesting but uncompetitive **Lotus 56B Turbine car,** but in 1972 Walker would be **Emerson Fittipaldi**'s teammate the entire season in the new **John Player Special** sponsored Lotus, a combination that looked great on paper. In the end, Walker's poor 1972 form defies explanation. While Emerson ran up front in all races, won five, plus finished on the podium in a few others, and won the championship with two races to go, the former king of Formula 3 did not score a single point. His best qualifying was 12th in Belgium, but more often than not, he qualified out of the top 20. His best finish was 10th in South Africa. early in the season. By Canada **Colin Chapman** had had enough of the Aussie and called former Lotus driver **Reine Wisell** to replace Walker. The Australian would get one more chance, for teams usually entered three cars in Watkins Glen in those days, due to good prize money, but Walker disappointed once more: he qualified 31 out of 32, retiring. Not surprisingly, he did not stir the interest of other team managers besides clueless **Maki** after such an awful display.

Trevor Taylor, 1963

David Walker was not the only Lotus driver to be utterly humiliated by a faster champion teammate. Trevor Taylor's 1963 season was just slightly better than the Australian. The Brit had already raced for Lotus in one race in 1961 and then the entire 1962 season, which started on the right path, with second place in Zandvoort. In 1963, Taylor also scored in the first race of the season at Monaco, one point from sixth. He could not imagine that would be the first and last point of a season where his teammate won seven GPs and the Championship, while his mount had all types of mechanical failures. That was just part of the problem, for even though Clark started most races from pole, Taylor's best start was a trio of 7th places, and a few times, he qualified out of the top 10. In non-championship races he did not do as bad, finishing 2nd in Pau and Kannonlopet, Sweden, but in one of such races, at Enna, he had a terrible accident that sidelined him from a championship round. The obvious result was dismissal from Lotus, and an equally inconclusive 1964 season at **BRP** and a final Grand Prix in 1966, driving the one-off **Shannon-Climax**.

In fairness to Trevor, his immediate successors at Lotus, **Peter Arundell** and **Mike Spence** did not fare much better. Arundell got two third places in 1964, but then an accident sidelined him for part of that season and the entire 1965 season. Spence replaced Arundell for part of 1964 and then 1965, but his performance was nowhere near Clark's or even Arundell's. Arundell came back in 1966 a shadow of his former self, and was replaced by **Graham Hill** in 1967.

Alex Zanardi, 1999

Just exactly what happened with **Alex Zanardi**'s season at **Williams** in 1999 will remain a question mark. A fast driver in Formula 3000, the Italian showed some pace driving for impecunious and sinking Lotus in its last legs in the 1993/94 F1 seasons, even scoring a point. Then he went to the United States, and three seasons in the CART Championship, driving for **Chip Ganassi**, showed a talented, skillful, assertive and surefooted driver who won many races and back-to-back championships in a very competitive series. For 1999 a glorious return to F1 appeared suiting to the Italian. Granted that in two short years Williams had fallen from status of obvious top team, and badly needed a works engine to remain on top. In the end, Zanardi did not score a single point, qualified midfield and the only place where he showed real pace was at Monza, where he qualified 4th. On the other hand, Zanardi's teammate **Ralf Schumacher** scored points in most races, had three podiums, often qualified in the top 10, got 35 points and finished 6th in the Championship. How is that for comparison? A chance lost, the Italian went back to the US racing scene, not knowing what hit him.

Ivan Capelli, 1992

There were many Italians in F1 in the late 80s, early 90s, and it was difficult to stand out in the group. Yet, Capelli did stand out, and some of his drives for **March/Leyton House** during the 1988 to 1991 are stuff of legend. Perhaps he lacked consistency, but the team was small, and despite having **Adrian Newey** as designer, could only do so much. After firing **Alain Prost** at the end of the season, who famously referred to the latest F1 offering from Maranello as a truck, Ferrari was looking for excitement for the Italian fans, and that came in the form of Capelli. The 92A was not a memorable car for a stretch, yet it was not terrible. While teammate **Jean Alesi** kept on trying, spinning off in the process, qualifying towards the front, and scoring a healthy 18 points including two podiums, Capelli seemed mostly disinterested, and his points tally at the end of the season was a meager 3. Not surprisingly, Capelli did not strike the fancy of other team managers after such a bad season, and a couple of races for **Jordan** were his last performances in the category.

Michael Andretti, 1993

Driving for **McLaren** in the early 90s was a dream. Having **Ayrton Senna** as a teammate, could be a nightmare. In the end, the nightmare prevailed for Michael Andretti, a very talented driver who was at the right place at the wrong time. By far the fastest active CART driver at the time, Michael had nothing to prove stateside, except winning the Indy 500. As for McLaren, in 1993 it was in damage control mood. After trying the **Lamborghini** engine in the off-season, the team settled on the 3**.5 Ford HB8** engine for the season, which was reasonable, but no match to the **Renault** engines used by **Williams**. Eager to keep Senna on board and sponsor **Marlboro** happy, the team worked around the clock to give the Brazilian at least a fighting chance. While Michael qualified in the top 10 in the first six races, accidents seemed to destroy his confidence little by little. From Canada on the American was qualifying in the second half of the grid, while Senna continued to score many points, always qualify in the top ten, and even get some wins. Michael's last race in Monza was truly melancholic. Once again he qualified in the top 10, this time doing well enough to finish 3rd and salvage a single podium from his sadly short GP career. So it is understandable that now he wants to be part of the show as a constructor: he does have unfinished business with the category.

Jan Magnussen, 1997

Magnussen was considered a true up and coming driver, amazing in Formula 3, having won no less than 14 races in the 1994 season. His debut driving a **McLaren-Mercedes** in the 1996 Pacific Grand Prix showed a balanced and careful driver, who qualified 12th and finished 10th. Not excellent, but not bad. When **Stewart** decided to pair the Dane with **Rubens Barrichello**, it sounded like a good idea: Rubens had been around since 1993, but did not seem as promising as the Scandinavian, whose innate speed would surely blossom. It was not to be. The only stellar qualifying in his 23-race Stewart history was 6th in Austria, 1997. Rubens got the best of him almost everywhere, in fact, while the Brazilian finished a fighting 2nd in **Monaco**, 1997, the Dane started last and finished 7th. In the end, Magnussen ended his F1 cycle earning a single point in the Canadian Grand Prix of 1998. Eventually Magnussen became a highly rated sports car driver.

Nelsinho Piquet, 2008-2009

Nelsinho Piquet, 3-time champion Nelson Piquet's son, was also highly rated. Very fast in Formula 3, the Brazilian was hired by **Renault**, in 2008, to partner **Fernando Alonso**. Nelsinho did well to lead a couple of races and finished 2nd in Germany, 2008. However, his involvement in the **Singapore** fiasco of 2008, where he was allegedly compelled to crash his car to favor teammate Alonso who had just pitted and ended up winning, did not make him any friends. The matter became public only after Nelsinho was fired from Renault having failed to score a single point in the incomplete 2009 season. Needless to say, he was never again seriously considered for a GP drive, and both team manager **Flavio Briatore** and engineer **Pat Symonds** were banned. Some sources claim he was considered for a **Ferrari** drive, but that seems very unlikely. Since then he has raced in many disciplines, and currently drives in the local Brazilian stock-car championship.

Ricardo Zonta, 1999-2005

Brazilians kept on looking for a new **Ayrton Senna**, and Ricardo Zonta looked the part. Ricardo won many championships, such as Formula 3, Formula 3000 and GT, but his Formula One career was inconclusive. From 1999 to 2005 he drove for **BAR, Jordan** and **Toyota**, scoring only 3 points and never looking competitive.

Bruno Giacomelli

Judging from his form in Formula 3 and Formula 2, Bruno Giacomelli looked like the future Italian hope. In F-2 alone he won 11 races between 1977 and 1978, and after being tried by **McLaren** in the 1977 Italian Grand Prix and in a few races in 1978, Bruno was hired by **Alfa Romeo** for 1979. He did well in the 1980 season, often qualifying in the top ten, specially towards the end of the season, and scored pole for the last **Watkins Glen** race. He led but dropped out.

While he could be trusted to place Alfa Romeos near the front of the grid on occasion during the 1981 and 1982 seasons, he often retired, but did get a single podium at Las Vegas, 1981. He moved on to **Toleman**, where he was utterly outclassed by **Derek Warwick**. Then Bruno vanished.

As a phoenix, he re-appeared in the 1990 season, driving for an overly ambitious project called **Life**, which, among other things, sported a proprietary W12 engine. Bruno never looked like close to pre-qualifying the beast in a busy 35-car entry, and not even changing to **Judd** engines for the last two races of the season resulted in a simulacre of success. He quietly left the F1 scene, never to return.

Derek Daly, 1982

If you look at the history of the sport very few drivers ever get a good chance to do well in the World Championship. Derek Daly did, and frankly, squandered it.

The Irishman was a fast driver in both Formula 3 and Formula 2, and his first cracks in Formula 1 were in less performing teams, such as **Ensign, March, Hesketh** and **Theodore**, or the once glorious **Tyrrell,** which was no longer a topline team when he raced there in 1980.

He began the 1982 season racing for Theodore, but then **Williams** ran into a problem as **Carlos Reutemann** quit and retired from Formula 1 after the Brazilian Grand Prix. **Mario Andretti** raced for the team in **Long Beach** but was not enticed to return to Grand Prix racing. As it was too late in the season to make any wonderful deals, Williams hired Daly as the number two driver.

Daly one race before his big break (Kurt Oblinger)

While it is true that the FW08 was not as good as the preceding car, **Keke Rosberg** scored constantly, won one race, had many second places, and at the end of season, was the champion. Daly, on the other hand, got three fifth places and two sixth places, scoring 36 points less than Rosberg. The difference between the teammates was tremendous, whereas in 1981 **Jones** and Reutemann were evenly matched.

Having blown such a chance, Daly was no longer considered for F1 berths, and decided to race in the US instead, where he competed in CART and then IMSA, ultimately winning races for **Nissan** in this category.

Mike Thackwell

In the late 70s Thackwell was the latest of a line of young New Zealanders to conquer the racing world in Europe. Like **Amon** and **McLaren** before him, Thackwell was a little more than an adolescent when he came to contest the Vandervell F3 championship in 1979, at 18 years of age. He acquainted himself rather well, won five times and in 1980 he became a Formula 2 driver. Then F1 chances appeared that very same season. Perhaps putting him in F1 cars that early was a bit mistimed, but Mike did practice an **Arrows** at **Zandvoort,** and then was entered by Tyrrell in **Montreal** and **Watkins Glen**. The results were disappointing. He had an accident in the first start in Canada and did not start the second time around and did not qualify at **Watkins Glen**. He continued in Formula 2 and became undisputed champion in 1984, driving a **Ralt Honda**. Thackwell would get two more chances in the category, in 1984, a **RAM** in Canada, where he retired, and then a **Tyrrell** in Germany, where he did not qualify. That was it, all improper entries for a driver of such talent. He then became a sports car driver, and before hitting 30, retired from the sport.

Others

Some highly successful Grand Prix drivers also had their off days. **Michael Schumacher's** return to F1 driving for **Mercedes** in 2010-2012 was not a good idea, in hindsight. **Ronnie Peterson's** season driving the 6-wheel **Tyrrell P34** in 1977 blotted his copybook somewhat, for it was generally felt the man could make anything go fast. **Jacky Ickx's** stint as a replacement for **Patrick Depailler** at **Ligier** in 1979 was a bad send-off for the talented Belgian. **Jody Scheckter's** 1980 season after winning the 1979 title was unbelievably bad. **Alan Jones`** brief returns to F1, first at **Arrows** and then at **Carl Haas'** team, are forgettable. **Nelson Piquet's** 1988 season with a turbo

Honda engine in the back was rather uncharacteristic. **James Hunt's** time at Wolf in 1979 was painful to watch. **Nigel Mansell's** weird foray at **McLaren** was puzzling to say the least. **Fernando Alonso's** 2015 season in a **McLaren-Honda** was awful. In more current times, **Sebastian Vettel's** current **Aston Martin** was a bad career end.

TEAM DNA

We have gotten used with the idea of Grand Prix teams being bought and changing names, which was not regular practice in the early years of the sport. There are many reasons for this, the most important is that setting up a modern grand prix team is a major undertaking, requiring rather specialized personnel, machinery, supply chains and locations. Of the last four GP teams to be formed from scratch, only **Haas** has showed the potential for lasting in the mid-term, and the 3 debuting teams in 2010 were just slightly better than the **Andrea Moda** and **Life** of yesteryear. While some of the name changes are just branding exercises, in some cases there has been outright change of ownership. The two most successful teams of the last ten years actually were created **Tyrrell** and **Stewart**. In a twist of fate, these two names that were quite intertwined in the 60s and 70s.

Tyrrell → BAR → Honda → Brawn → Mercedes

Tyrrell's success came mostly in its first four years of existence, while **Jackie Stewart** drove for the team. The constructor got a pole position on debut, then won 7 races out of 11 and the 1971 Championship. Stewart would be runner-up in 1972, and then won the title again in 1973, in the process winning fifteen races in his Tyrrell years. As the Scot retired at the end of the season, the British team never again featured as a dominating figure. Wins after the Stewart years were few, in spite of the efforts of **Jody Scheckter, Patrick Depailler, Michele Alboreto** and **Jean Ales**i. Time was ripe to leave the sport in the late 90s, and good ole' Ken sold the team to **BAR**, who signed 1997 World Champion **Jacques Villeneuve,** no less. F1 is full of very optimistic dreamers, and **Craig Pollock** fit that category, so the early BAR years were dreadful. As **Honda** came on board, by 2004 **Jenson Button** had managed to finish 3rd in the championship, and BAR became more and more a Honda works team. The change was formalized in 2006, and Honda got its first works win since the 1967 Italian Grand Prix at Hungary. Then the 2008 **Smith Barney** saga became the tip of the iceberg of a scary global economic collapse, and Honda quickly became one of the three works teams to flee from the expensive sport in a couple of years, the other two being **BMW** and **Toyota**. It was a pity, for the 2009 car appeared promising after a terrible 2008 season and simply closing the team would be a waste, so the outfit was basically transferred over to **Ross Brawn,** sold in a sweetheart deal. The new Brawn team is perhaps the most amazing story in F1. Equipped with **Mercedes** engines, the novice team won six of the first seven races, plus two others, and drivers **Jenson Button** and **Rubens Barrichello** (who drove together in the Honda team) got the first and third places in the championship. The rest is history. Mercedes bought Brawn, and after a first few learning curve years, dominated F1 from 2014 to 2020, and it only shows signs of losing stamina now.

Tyrrell was the first team to have Benetton sponsorship in F1. Eventually the company became a constructor on its own (Kurt Oblinger)

Stewart → Jaguar → Red Bull

Jackie Stewart was **Ford's** main man in Formula 1, in fact, the only one to be paid by Ford, and remained a brand ambassador after retiring. Son **Paul Stewart** tried his hand at driving, but lacked the skills, charisma and health of his father, but did set up quite a good lower formulae team. In 2007, with Ford's substantial assistance, the **Stewart Formula One** team was formed, and although it was never a top team in its short existence, it did win a race with **Johnny Herbert** and was a regular points scorer in 1999. Ford was reasonably convinced that buying the team from the shrewd Scot and naming it **Jaguar** would be great for its international premium brand. Truth is, Jaguar looked like a sinking boat from the start. Management changes and corporate meddling hurt it, and hiring **Bobby Rahal** who had scant F1 experience seemed a bad start. Changes of personnel ensued, and by 2004 the good idea seemed a bad one. The team was sold to **Red Bull** for the 2005 season, which became a powerhouse in the category since then.

Jordan → Midland → Spyker → Force India → Racing Point → Aston Martin

Jordan had a very nice car in its 1991 debut season, and became known as the team that gave **Michael Schumacher** its debut but the association did not last long, as the German was hired by **Benetton** right away. After changing engine suppliers a few times, Jordan finally posted its first victory with **Damon Hill,** in 1998, followed by two wins in 1999, driven by **Frentzen**. The first change of ownership took place in 2006, as the **Midland** team emerged. That quickly changed into **Spyker,** in 2007, which is actually a Dutch sports car manufacturer, and managed to lead a race under freakish conditions in Germany. From Irish, to Russian, to Dutch to Indian hands, the team became **Force India** in 2008. The small team had spirited drivers through the years, such as **Perez, Hülkenberg, Sutil, Fisichella**, but never won a race. In the middle of the 2018 season, Canadian billionaire **Lawrence Stroll** bought the team and changed its name to **Racing Point**. The team's moment of glory came in 2020, as Mexican driver **Sergio Perez** finished the championship in fourth place and won the Sahkir Grand Prix in **Bahrain**. Since then Stroll had also acquired control of Aston Martin, and adopted the name for the team, hiring former 4-time champion Vettel for the 2021 season. So far Aston Martin has failed to impress.

Toleman → **Benetton** → **Renault** → **Lotus F1** → **Renault** → **Alpine**

Toleman stayed for a short while in F1, quickly becoming Benetton (Kurt Oblinger)

Toleman was a successful F-2 team/constructor which entered the F1 wars in 1981 bravely carrying down on power **Hart** turbo engines. The debut season was dreadful, but eventually Toleman became a midfield team. In 1986 sponsor **Benetton** bought the team outright and won in the debut year. Eventually Benetton became a top team, won a couple of championships with **Schumacher** and stayed on a few more seasons. Engine supplier **Renault** eventually bought out Benetton, rebranded the cars Renault, and won a couple of titles in 2005/2006. Long-term success was not sustained, and in 2011 Renault became **Lotus F1** winning a couple of races with **Kimi Räikönnen** in 2012 and 2013. As the tie-up with **Red Bull** went sour, Renault decided to rebrand bought the team back from Lotus F1 in 2015, renamed it Renault, and in 2021 changed identity once more, adopting the name **Alpine**, itself a revived road car brand.

Sauber → BMW → Sauber → Alfa Romeo

Sauber had been a race car constructor since 1970, but until the early 90s it built only sports cars. After becoming **Mercedes**' chassis partner in Group C, the Swiss outfit was chosen to reintroduce the German manufacturer into F1. The tie up did not last long, soon Mercedes decided **McLaren** had better chances of success, yet Sauber stayed around, mostly on the strength of **Red Bull** sponsorship. Then Red Bull also left Sauber, to build its own team! Talk about disappointment. BMW came to the rescue, for a while at least. The new BMW-Sauber team soon became competitive, **Robert Kubica** won a race for the team, but then the 2008 economic meltdown came and BMW took a walk and left F1 again. Another let down. Sauber became Sauber again, until, in 2018, the team became known as **Alfa Romeo**. Let us see how long this will last…

Minardi → Toro Rosso → Alpha Tauri

Minardi started in F1 in the mid 80s, powered by **Chiti's Motori Moderni** engines. It never won a GP but gave a large number of drivers a chance to start a F1 career, including future world champion **Alonso** and **Webber**. It did build up a lot of goodwill with time, and it was everyone's second favorite, it seemed. However, by 2005 it became too much for the current owner **Paul Stoddart** to remain in business so he sold the team to deep pocketed **Red Bull**, essentially becoming Red Bull's B Team. The renamed Toro Rosso changed name to Alpha Tauri in the 2020 season, named after Red Bull's budding fashion brand. Minardi never won anything, but both Toro Rosso and Alpha Tauri have won a race apiece, to their credit.

Rondel → Token → Safir

I know, the Rondel F1 was a planned team that never happened. **Ron Dennis**, who built his own F-2 cars in the 1973 season, decided to enter F1 in 1974, with a car designed by **Ray Jessop**, but wisely decided the time was not right to enter the category after losing **Motul** support. When it saw light of the day, the car came out as the **Token**, one of the 8 new constructors that debuted in F-1 in 1974. It did a few races and fell into oblivion. In 1975 it reappeared in non-championship F1 renamed Safir. It only makes the list because the changes took place within a couple of seasons.

Politoys → Iso-Marlboro → Williams → Wolf-Williams → Wolf

Frank Williams had been entering outsourced cars in the F1 Championship since 1969, under the banner of Frank Williams Racing Cars. After running **Brabham, De Tomaso** and **March**, he decided to build his own car in 1971, which was only finished in 1972, and named Politoys. This was the first F1 car to be named after a sponsor, an Italian manufacturer of toys and model cars, even though the **Lotus** at the time was often referred as **John Player Special**. The Politoys raced just a few times in the season, and did not impress. For 1973 Frank got new sponsors, **Iso Rivolta**, an Italian maker of luxury and sports cars, and **Marlboro**. The cars named Iso-Marlboro raced in the 1973 and 1974 seasons, but despite the flashy sponsorship, Iso Rivolta went into liquidation during the troubled financial times of 1974, and payment of sponsorship from that corner was an obvious low priority on the mind of the liquidator.

The Iso-Marlboro in happier days, early 1974 (Alejandro de Brito)

Frank then gave up on the idea of naming cars after sponsors, and his cars became Williams in 1975. For 1976 Frank entered into an agreement with **Walter Wolf**, a Canadian millionaire, and the newest Hesketh design was bought from **Lord Hesketh**, who was withdrawing his team. With **Jacky Ickx** and designer **Harvey Postlethwaite** on board, things looked auspicious for Frank, but it turned out bad as the cars looked better than they performed. During the season Wolf bought Frank Wllliams's share of the team, which became **Walter Wolf Racing**, while the cars continued to be run as Williams for the rest of the 1976 season, unsuccessfully. For 1977 Wolf changed the name of his team's cars to Wolf, which had a wonderful debut season winning three races and finishing the driver's championship in the runner-up position with

Jody Scheckter. The next two seasons were not good, Scheckter had a subdued performance in 1978, and new hiring **James Hunt** disappointed immensely in the early part of 1979, retiring mid-season. At the end of the season Wolf decided to spend his money elsewhere, closed the team and sold the equipment to **Fittipaldi**.

As for Williams, he started **Williams Grand Prix Engineering** in 1977, first running a **March** for **Patrick Neve**, then building **Patrick Head's** designs starting in 1978. In short order, helped by plentiful Saudi money, Williams became the class of the field and a force to be reckoned with for a couple of decades.

Note that these two Williams teams are not connected, they are two separate entities which happen to share the name of the owner.

The FW06 was Williams' first proprietary chassis in the Williams Grand Prix Engineering era (Alejandro de Brito)

Teams that rebranded the car's name in deference to sponsors, without change of ownership:

March → Leyton House → March (1990 to 1991)

Arrows → Footwork → Arrows (1991 to 1996)

Arrows raced in F-1 from 1978 to 2002, and never won a race, although it led its second race. (Alejandro de Brito)

Teams that rebranded car awarded by a Court in a legal proceeding.

Ensign → Boro (1976 to 1977)

Teams that rebranded car due to change of ownership but closed soon after

Onyx → Monteverdi (1989)

DYNASTIES

Quite a few relative pairs have raced in Formula 1, some of them concurrently. This not atypical in any sport, although some dislike name dropping as a means of getting ahead in F1. The high level of competition in this field of motor racing, where merit should count more so than family ties, is the reasoning for such dislike. But let's face it: having a famous father or brother does open doors, which is specially true of such a media driven discipline as F1.

FATHER/SON

Jos/Max VERSTAPPEN

Father Jos had a great chance in the 1994 **Benetton** team, as number 2 driver **J.J. Lehto** was unable to race in the first events of the season, so he was called to fill-in. Jos did not disgrace himself with 9th on the grid, but his performance was nowhere near brilliant **Michael Schumacher's**. While Lehto was back for the San Marino Grand Prix, he proved to be slower than expected, so Verstappen was brought back to the team in Britain. The Dutch driver got a couple of thirds, to finish the year on good terms with Benetton but was not retained. Jos' career never really took off in the category. After stints in **Simtek, Tyrrell, Stewart, Arrows** and **Minardi**, he disappeared briefly from the GP circus in 2003. Twelve years later he was back, this time as mentor, coach and manager for his son Max, whose career was successful from the get-go. Max' career was another story. Brought to F1 at a very young age as a **Toro Rosso** driver, a little over a year latter Max was moved over to main team **Red Bull** and won in his debut there. He has remained in the team and won the 2021 title under controversial conditions and seems on path to winning his second straight title.
Jos = 0 wins, Max = 1 title, ongoing career

Gilles/Jacques VILLENEUVE

Gilles rose to prominence in mid 70s North American Formula Atlantic, which led to a ride in the **Wolf** Can-Am team, whose manager **Chris Amon** recommended the driver to Enzo Ferrari. After debuting for **McLaren** in 1977, Gilles was hired by the Scuderia at the end of the season as a replacement for the departing **Niki Lauda**. Gilles career in F1 was brief, his wins few, but many still remember his speed and outstanding car control, learned in snowmobiles. Dying an untimely death in 1982, it would take until 1996 for this side of the Villeneuve family to be represented in F1. **Jacques,** the son, had a good run in lower Formulae, but is best remembered for his 1995 Indy 500 and CART title wins. He was quickly hired as a replacement for the departing **Coulthard** at **Williams**, and the junior Villeneuve was quick right off the bat. Jacques could do no wrong for a couple of years and was runner-up to **Damon Hill** in 1996 and Champion in 1997.
Unfortunately, this form was not sustainable, and Jacques' career took a dive in latter years, until he faded for good in 2006.
Gilles = 6 wins, Jacques = 1 title, 11 wins

Graham/Damon HILL

Graham was a very popular figure and sports ambassador, who started late. Notwithstanding, Graham was world champion twice, in 1962 and 1968, an **Indy 500** winner in 1966 and **24 Hours of Le Mans** winner in 1972, and widely acknowledged as **Jim Clark's** main challenger during the 1.5-liter years. Son Damon also started late in car racing, at first racing bikes. Whereas Graham's accomplishments are widely recognized, Damon remains to this day an underrated driver, who did outstanding work while driving for **Williams** in the 1993 to 1996 period. He proved his worth leading a GP in a weak **Arrows-Yamaha** in 1997 and then winning the first GP for **Jordan** in 1998. The fact is that Damon had 22 wins from 115 races, to his father's 14 wins in 175.

Graham = 2 titles, 14 wins; Damon = 1 title, 22 wins

Graham Hill envisioned a future as team owner and constructor which was cut short in the end of 1975. Here are his 1974 Lolas (Alejandro de Brito)

Jack/David BRABHAM and Jack/Gary BRABHAM

Jack had an outstanding career in Formula 1, becoming the second 3-title winner in 1966, doing so in a car of his own construction. Additionally Brabham was the first driver to win the world title driving a rear engined car, a **Cooper** in 1959, and was still very competitive in his last season, 1970, at 44 years old. His three sons had some achievement in motor racing. **Geoff** never raced in F1, but won Le Mans and multiple IMSA Championships. **David** became a top sports car champion and **Gary** was a top British Formula 3000 driver. However, none of his sons shone in F1, although none had a proper chance. David drove for the **Brabham** team in 1991, without great results. Jack = 3 titles, 14 wins; David = 0 titles, 0 wins; Gary = 0 titles, 0 wins

Hans Sr/Hans Joachim STUCK

Hans the father was a major figure in 30s racing, known as a common fixture of the Auto Union team from 1934 to 1939, who won many GPs and a large number of hill climbs. He drove until his early 60s, and drove in a few occasions in the World Championship. As for Junior, he was considered a possible GP winner when he came into Formula 1 in 1974, but never had the proper opportunity. He did lead the 1977 U.S. East Grand Prix, under pouring rain, and became a main factor in sports car racing.
Hans Father = 0 wins, Hans Junior = 0 wins

Hans Joachim Stuck raced in Formula 1 from 1974 to 1979. Here was his last ride, the ATS. (Kurt Oblinger)

Jan/Kevin MAGNUSSEN

Jan was highly touted a possible great when he came into F1 in 1996, having won a huge number of F3 races in 1994. Initially a **McLaren** driver, due to his **Mercedes**

connections, in 1997 Jan went to **Stewart**. The relationship did not prosper; Jan proved much inferior to teammate **Barrichello** and would not last the 1998 season. He did become a top sports car and GT driver, winning many events and racing to this day. Kevin's career began the best possible way, as he finished his debut race in 2nd place driving a McLaren in Australia, 2014. That was, unfortunately, the highest point of his career. After an inconclusive season at **Renault,** 2016, Kevin became a **Haas** driver, where he had some good moments, some bad ones. After sitting out 2021, Kevin was brought back into the team, in 2022, a popular move. He has developed into a very sound performer, although not world champion material.
Jan = 0 titles, 0 wins; Kevin = ongoing driver

Mario/Michael ANDRETTI

Mario drove in F1 from 1968 until 1982 and had a huge impact on the sport. He scored pole on his very first race, the U.S. Grand Prix of 1968, and won his first GP in 1971, driving for **Ferrari**. His best years in the category were 1976 to 1978, when he won most of his races and was a deserving world champion. He was obviously the class of the field in the 1977 and 1978 seasons and continued to race into the mid 90s. Michael was a major disappointment in Formula 1, there is no other way to describe his 1993 season. A number 2 to **Ayrton Senna** at **McLaren**, Michael felt out of place in F1, crashed often, and could not get nowhere near Senna's speed. He did have a podium in Monza (3rd) place, before leaving and never coming back. Now a very successful team owner in Indycars and elsewhere, he has been trying to enter Formula 1 as a constructor. An attempt to buy out the **Alfa Romeo** team in 2021 failed, and he now feels there is some resistance to his entering the category.
Mario = 1 title, 12 wins; Michael = 0 title, 0 wins.

Mario Andretti in 1978 (Alejandro de Brito)

Michael/Mick SCHUMACHER

What can one say about Michael? Some felt Germany's best hope in the late 80s - early 90s was **Heinz-Harald Frentzen**, but Michael dispelled any doubts by becoming a quick learner after a debut in **Jordan** and a move to **Benetton** in 1991. He stayed in the latter from 1992 to 1995, becoming a top driver and scoring many points in the process. At first wins were few, but by 1994 Michael had become the top driver, clearly beating **Ayrton Senna** in the first two races of season, then winning the title for two straight years. He moved to **Ferrari** in 1996 but it took a while to achieve the desired prize, a driver's title, which the Italian team had not won since a very far 1979. That finally came in the year 2000, and then Michael simply dominated the sport, winning five straight titles. A return to racing at **Mercedes** for three seasons was inconclusive, and a skiing accident after the second retirement has left the great driver far from the public's eyes, under unknown conditions. As for Mick, expecting him to be the next Michael is a bit too much. He did have a terrible first season in 2021, driving a bad Haas, but has performed

admirably well in the 2022. If he ever gets the promised Ferrari seat, who knows, a few wins may be in order.
Michael = 7 titles, 91 wins ; Mick = ongoing career

Wilson/Christian FITTIPALDI

Fittipaldi became a household name in the sport as **Emerson** won his 4th race in the category, the U.S. Grand Prix of 1970. His older brother Wilson had actually tried Formula 3 in Europe, 1966, but went back to Brazil disappointed that the promised support did not materialize. Wilson came back to do F3 in Europe, 1970, in the wake of his brother's success in the category, debuted in Formula 1 in the non-championship Argentine GP of 1971, and got a Brabham seat in 1972. He scored three points in 1973 and seemed to be on the way to a podium at **Monaco** that same season. In late 1973 he announced the start of the Fittipaldi F1 team, taking a year off to build the team and the car, and returned for a last season of F1 racing in 1975, scoring no points. His son Christian did race in Formula 1 briefly, driving for **Minardi** and **Footwork**, 1992 to 1994, and to his credit scored points in all seasons, however did not do enough to impress team managers in the top teams. He wisely decided to move over to CART in the USA.
Wilson = 0 titles, 0 wins; Christian = 0 titles, 0 wins

Teddy Pilette suffered trying to qualify the BRM P207 in 1977 (Teddy Pilette collection)

Andre/Teddy PILETTE

Andre had one of the longest career spans in Formula 1, from 1951 to 1964. However, the Belgian, who was the son of a driver from the formative years of the sport, never drove a full season and except for a chance at **Ferrari**, in the 1956 Belgian Grand Prix, never drove top machinery. He did get two points for his troubles. Son Teddy was touted for a seat in the Parnell team in the early 60s, but that did not materialize. He did get his first chance after a successful career in sports cars and Formula 5000, driving a Finotto **Brabham**, in the 1974 Belgian Grand Prix. That ended up being his only career start, as his next chances came in the slow **BRM P207**, which he failed to qualify wherever he tried, in 1977.
Andre = 0 titles, 0 wins; Teddy = 0 titles, 0 wins

Manfred/Markus WINKELHOCK

Manfred was a successful sports car driver, associated with both **BMW** and **Ford**, who drove occasionally in Formula One from 1980 to 1985. His first outing, at **Arrows**, in the 1980 Italian GP was a disappointing DNQ. He was hired by the unstable **ATS** team in 1982, managing to stay on board for three seasons. He scored 2 points from 5th after the top 2 drivers were disqualified in Brazil, 1982, which was his second race for the team but did not score any further. He did put the ATS in the top 10 of the grid many times in the 1983 and 1984 seasons, after the team got BMW engines, but did not translate that into race results. A move to **RAM** in 1985 did not yield any results, and he died in a sports car race in 1985. His son Markus had the distinction of leading his single Formula One race, at the European Grand Prix, 2007, that time held at the Nürburgring. Driving a slow **Spyker-Ferrari**, Markus, who was replacing **Christijan Albers** and started last, pitted for wet tires earlier than everybody in the changing conditions,

and shod with the right tire choice, built up a lead of over 30 seconds. He led for six laps, but the car retired with hydraulic issues.
Manfred = 0 titles, 0 wins; Markus = 0 titles, 0 wins

Keke/Nico ROSBERG

Here is another set of son/child champions. Again, the son is highly underrated, a sad state of affairs, but to some extent, so is the dad. Keke was a major factor in Formula Super Vee when the entered Formula 2 in 1976. He was soon racing all over, until reaching Formula One in 1978, hired by the **Theodore** team. He actually won a non-championship Formula One race at **Silverstone**, run under deluge conditions, but the form in that car was not repeated. Further drives for **ATS**, driving a private **Wolf** and a works Wolf did not yield any points, but the Finn got a podium from his first outing at **Fittipaldi**, in 1980. This turned out to be his top performance in two years in the team, going scoreless in 1981. A big chance came in 1982, as both **Alan Jones** and **Carlos Reutemann** walked away from excellent **Williams** drives. At first subdued, little by little Keke scored enough points to pass Pironi, and ended up champion, the only one to do so without scoring a single point in the preceding season. Some felt Rosberg's title was undeserving because he won a single race, forgetting that no less than 11 drivers won races that year, and no one won more than two races, anyhow. Rosberg would remain in F1 only a few more seasons, quitting after 1986. Detractors should consider that Rosberg raced 512 laps in the lead in F-1 out of 114 starts, more laps than **Emerson Fittipaldi, Denis Hulme, Jacques Laffite and John Watson**, among others. Nico came into F1 in 2006, driving for Williams. A former GP2 champion, Nico scored fastest lap in his first GP, and remained at Didcot until the 2009 season. For 2010, he moved to **Mercedes**, where he had success. I suppose Nico's major accomplishment was being the only teammate to consistently outdrive **Michael Schumacher** in

F1, and he kept **Lewis Hamilton** on his toes until retiring a worthy champion in 2016.
Keke = 1 title, 5 wins; Nico = 1 title, 23 wins

Keke Rosberg actually managed to win a F1 race with the slow Theodore in 1978, although a non-championship event (Alejandro de Brito)

Jonathan/Jolyon PALMER

Dr. Jonathan is best remembered for being a medical doctor who raced in Formula One, rather than from his feats in the category. One of the last European Formula 2 champions, in 1983, Palmer drove in F1 from 1983 to 1989. His first race was in a **Williams,** the first and last one for a top team. He then drove one season for **RAM,** two for **Zakspeed,** and three for **Tyrrell**. In the latter he won the Jim Clark Trophy for normally aspirated cars, his best overall result a 4th. In the next two seasons he got another 7 points, so, in total 14 points from 82 starts. He did get a fastest lap in Canada, 1989. After retirement he became known for being the promoter of the **Palmer Audi** racing series. His son Jolyon, who had won the GP2 series in 2014, spent two years driving for the **Renault** Formula One Team, in 2016 and 2017, and his best result was 6th in

Singapore. He seemed out of his depth in the category and left never to return after the 2017 season.

Jonathan = 0 title, 0 wins; Jolyon = 0 title, 0 wins

Nelson/Nelsinho PIQUET

Nelson Piquet the father had an outstanding F1 career, not matched by his son (Kurt Oblinger)

Nelson Piquet's overall numbers stack up well against the hundreds of drivers who have competed in the World Championship since 1950, with many wins, poles, podiums, fastest laps and laps in the lead, in addition to 3 championships won in 1981, 1983 and 1987. Diehard Nelson fans believe he is not properly appreciated, and there is a degree of truth to that. It also true that Nelson has had a penchant for controversy since his Formula 3 days, and has had mostly a hate-hate relationship with the press, which might explain the lack of appreciation. Son Nelsinho's performance in lower formulae was excellent, so much was expected of him when he was hired by **Renault** for the 2008 season. He did well enough in his maiden season to finish a race in second, lead laps and score points. In 2009 he scored no points and was sacked

by Renault before the end of the season. Then came Singaporegate, reported in the section Disappointments, and that was the end of his career in F1.
Nelson = 3 titles, 23 wins; Nelsinho = 0 titles, 0 wins.

Satoru/Kazuki NAKAJIMA

Satoru's greatest feat was becoming the first permanent Japanese driver in Formula 1, largely with the support of **Honda**. Satoru managed to stay in F1 from 1987 to 1991, driving Honda powered cars in three of these seasons. He was never a top line driver, and his best result was a couple of 4th places in the 1987 British Grand Prix and the 1989 Australian Grand Prix. He qualified in the top 10 five times in 1988, when he had one of the few turbo cars in the field but was mostly a midfield qualifier elsewhere. He did score the fastest lap in the 1989 Australian race. His son Kazuki was a **Toyota** man, which served him well both on and off F1. In Grand Prix racing he drove for Williams, which had Toyota engines between 2007 to 2009. His best result was earned in his second race, a 6th in Australia, 2008. After leaving F1, Kazuki became a sports car driver for Toyota, and won the **24 Hours of Le Mans** three straight times, having since retired from the sport.
Satoru = 0 titles, 0 wins; Kazuki = 0 titles, 0 wins

Chanoch/Roy NISSANY

Until the date of writing no Middle Eastern driver had ever started a Formula 1 race, although there are four GPs in the region now. However, the Israeli father/son Nissany duo have tried to break into the sport but have the distinction of being the least successful relative pair. Usually at least one of the relatives has had a successful career, in or out of F1, which does not apply to either Nissany. Chanoch tested for **Jordan** in 2004, which lead to more testing in **Minardi** in the 2005 season. The team invited the 42-year old to be the third driver in the

Hungarian Grand Prix, and his performance can only termed an embarrassment. Son Roy got his chance in 2020, as a third driver for the **Williams** team. He ran as the third driver in six occasions and did not impress. One should also remember that the 2020 Williams was a rather bad car to begin with.
Chanoch: 0 titles, 0 wins; Roy: 0 titles, 0 wins

SIBLINGS

Jackie/Jimmy STEWART

Jackie was a major driver from the mid 60s to the early 70s, and for a while held the record for most Grand Prix wins, at 27. The Scot followed on the path of **Jim Clark**, however managing to come out of the sport alive. Stewart was also one of the first drivers to seriously pursue safety in Formula One, which earned some criticism from some corners. He is also reputed to be the first driver ever to earn more than one million dollars in a season, 1971, and retired relatively early at 34. He remained a sports ambassador to this day and ran a Grand Prix team under his name from 1997 to 1999. Brother Jimmy, on the other hand, had a very subdued career and his only GP start was at Britain, 1953.
Jackie = 3 titles, 27 wins; Jimmy = 0 titles, 0 wins

Jackie Stewart in a Tyrrell, 1972 (Russell Whitworth)

Michael/Ralf SCHUMACHER

Following the footsteps of Michael Schumacher must have been very hard for Ralf, but in hindsight, he did rather well. The Schumacher brothers were, by far, the most successful pair ever, and Ralf's career span went only one season beyond his big brother's first time retirement, ending in 2007. The level of success between the brothers cannot be reasonably compared but Ralf did win six races, more than a few champions, and had a fair share of poles and fastest laps. Additionally, Ralf mostly raced for the **Williams** team past its best, and **Toyota,** his last team, was never very competitive.
Michael = 7 titles, 91 wins; Ralf = no titles, 6 wins

Jody/Ian SCHECKTER

Jody came into F-1 a very young 22, in 1972, and impressed right away, even though he was prone to crash and create havoc during the 1973 season. He did survive his wild season and after joining Tyrrell in 1974 he became a neater and less spectacular driver, but effective enough to end his maiden season with the possibility of winning the title. A move to **Wolf** in 1977 produced three wins and a runner-up position in the championship, and the expected title did come in 1979, at **Ferrari.** He had a very poor season in 1980 and quit in the end of the year. Older brother Ian entered F-1 in 1974, driving a Team Gunston **Lotus** in his home race in South Africa, and also got drives in **Hesketh** and **Williams** in 74 and 75. He only had a single full season in F1, in 1977, when he raced for **March**. Unfortunately, the March works team was on the way out at the end of the season, and the cars were slow all year, although Ian qualified the car eighteenth out of 30 in Germany. He then went back to South Africa, where he dominated Formula Atlantic for many years.
Jody = 1 title, 10 wins; Ian = 0 titles, 0 wins

Jody Scheckter early in his career. Love the sideburns. (Rob Neuzel)

Manfred/Joachim WINKELHOCK

Not only did Manfred's son eventually race in F1, very briefly, it turned out, but his brother Joachim also tried the category. Unfortunately for him, the chance came in the oversubscribed season of 1989, when as many as 39 cars fought for a place under the sun. Joachim happened to race for **AGS**, one of the worst teams around, and did not even pre-qualify the seven times he tried. Notwithstanding, Joachim was a champion in many disciplines and also won the **24 Hours of Le Mans**.
Manfred = 0 titles, 0 wins; Joachim = 0 titles, 0 wins.

Emerson/Wilson FITTIPALDI

Here is another pair of brothers that raced concomitantly in Formula One, with different fortunes. While younger

brother Emerson quickly achieved success in the category, racing for **Lotus** and **McLaren** and winning many races and two championships, Wilson struggled somewhat to find pace, but had he remained at **Brabham** in 1974, perhaps the BT44 could have proved a better mount than the BT33, BT34 and BT42 he drove before. He stayed off the tracks in 1974 to prepare the **Fittipaldi** F1 challenger, which sadly, did not achieve the expected success, and turned Emerson into a mid-field runner at best. Eventually Emerson retired from F-1 and the Fittipaldi team stayed around another two seasons, without proper sponsor support. The end came in 1982, but the Fittipaldi name would come back later...
Emerson = 2 titles, 14 wins; Wilson = 0 titles, 0 wins

Ernesto/Vittorio BRAMBILLA

The Brambilla brothers raced at different times in Formula One. Ernesto, or Tino, had a couple of attempts in the big time, first in a Centro Sud **Cooper** in the 1963 Italian Grand Prix, for which he did not qualify, and then a **Ferrari** in the 1969 Italian Grand Prix, but the car was taken over by **Pedro Rodriguez**. Younger brother came into F-1 much later than Ernesto but had greater success. Vittorio managed to stay in F1 from 1974 to 1980 and did a bit of everything: won a half-race, got a pole, a fastest lap, led races, scored a few points and crashed a whole lot. The 70s were not kind to Italians in Formula 1, and the fact that Vittorio was the most successful of the lot tells the story.
Vittorio = 0 titles, 1 win; Ernesto = 0 titles, 0 wins.

Gilles/Jacques VILLENEUVE Sr.

I have discussed Gilles' career above, in connection with son Jacques. It so happens he also had a brother called Jacques, so this is not a typo. Brother Jacques, as opposed to son Jacques, was much less successful than Gilles in Formula 1 and elsewhere. He was older than Gilles, but his first F1 attempt was in the Canadian Grand

Prix of 1981, where he failed to qualify an **Arrows**. He also failed to qualify in Las Vegas, that same season. In 1983, he again tried to qualify a car in the Canadian Grand Prix, this time a **March RAM**, and also failed. That sums it up. The highest point of his career was winning a CART race in 1985 and also the 1983 Can-Am title, by that time quite a minor series.

Gilles = 0 titles, 6 wins; Jacques = 0 titles, 0 wins

Ricardo/Pedro RODRIGUEZ

A very young Ricardo debuted for **Ferrari** in the 1961 Italian Grand Prix, surprising all in attendance by posting the 2nd fastest time in qualifying. In the race things did not go so well, and he retired. Ricardo would race four times during the 1962 season and got a 4th in Belgium and 6th in Germany. He met his death at 20 years of age driving a Lotus in practice for a non-championship Mexican Grand Prix, and was a very fast driver, widely thought of as a future world champion. Older brother Pedro raced in Formula 1 for many years, from 1963 to 1971, driving for **Lotus, Ferrari, Cooper** and **BRM**, and won his first Grand Prix in South Africa, 1967. A winner at **Le Mans**, 1968, he became the fastest sports car driver in the world, and his skills driving the **Porsche 917** are legendary. He would win a second Grand Prix in 1970, in Belgium, driving for **BRM**, but died driving a **Ferrari 512** in an Interserie race in the 1971 season.

Ricardo = 0 titles, 0 wins, Pedro = 0 titles, 2 wins

GRANDFATHER/GRANDSON

Emerson/Pietro FITTIPALDI

Pietro is the first grandson of a former championship driver to race in F1. Holding dual citizenship, Pietro runs under the Brazilian flag, although he was born in the USA. This association has not helped Pietro much. He has been racing cars since 2011, in many disciplines, from

NASCAR, to F3, DTM, Sports Cars, Indycars, etc. He became a reserve driver for the **Haas** team in 2019, and had a couple of chances in 2020, replacing the injured **Romain Grosjean**. The big break seemed to come in 2022, as Haas severed ties with Russian driver **Nikita Mazepin**, and the drive seemed to be pretty much his. Sadly for Fittipaldi family fans worldwide, Haas decided to call back **Kevin Magnussen**, and worst yet, the car is not bad at all.

Emerson: 2 titles, 14 wins; Pietro: 0 titles, ongoing career

Emerson Fittipaldi's best season in the Fittipaldi was 1978 (Alejandro de Brito)

UNCLE/NEPHEW

Ayrton/Bruno SENNA

Ayrton's legend only grew in the aftermath of his tragic death at **Imola**, 1994. He had finally gotten his hands on a **Williams-Renault**, but then the car broke at Tamburello, killing the driver and bringing F1 into turmoil. His nephew Bruno made it to Formula 1 in 2010, as a driver for **Hispania(HRT)**. He did the full season, and his best result was 14th with a best qualifying 18th in Belgium. The

relationship with the team was strained, and by 2011 Bruno found his way into the **Renault** team, a much better team, replacing **Nick Heidfeld**. There his best result was a 9th, meaning 2 points. Next was an emotional move to **Williams** in 2012 the team for which his uncle raced at the time of his death, where he scored points in ten occasions, including a fastest lap in Belgium. That was not sufficient to entice anyone to give the Brazilian further rides in F1, so that is the last we have heard of the Senna last name, for now.
Ayrton: 3 titles, 41 wins; Bruno: 0 titles, 0 wins.

Jo/Jean-Louis SCHLESSER

Frenchman Jo had long experience when he first raced a proper F1 car in the World Championship, for he started racing in 1952, and competed in a number of categories including rallying. Jo was entered in the F2 section of the German Championship, in 1966 and 1967, driving **Matras**. He acquainted himself well, and then was invited to drive the brand-new air-cooled **Honda** 3.0 liter car in the French Grand Prix of 1968. It rained a lot at Rouen that day, and Jo lost control of the car, which caught fire very quickly, given the amount of manganese in the chassis, killing the 40-year old driver. His nephew Jean-Louis had climbed the ladder as most French drivers of the 70s and 80s did. He tried in his hand in F1 perhaps a bit early, in the French GP of 1983, and did not qualify a RAM **March**.

A sketch of the Honda RA302 that killed Schlesser

He found success as a top sports car driver for the **Sauber-Mercedes** team, winning many races and championships, when a golden chance appeared: he was invited to replace the injured **Mansell** in the **Williams** in the 1988 Italian Grand Prix. Dream turned into a nightmare, as Jean-Louis qualified poorly, then tangled with **Ayrton Senna** who was about to lap him in the closing stages of the race, causing the Brazilian's retirement. Jean-Louis still finished 11th, and that was his last chance in F1.
Jo - 0 titles, 0 wins; Jean-Louis: 0 titles, 0 wins.

GRAND UNCLE/GRAND NEPHEW
Lucien/Jules BIANCHI

Belgian Lucien Bianchi was competitive in sports cars and other categories and managed to stay relevant in F1 between 1959 to 1968. He scored a point in his very second Grand Prix, driving a **Cooper** for Equipe Nationale Belge in his country's Grand Prix of 1960, finishing 6th, but until 1968 he did not have a steady GP ride. That finally came up in 1968, as Lucien was hired to drive the heavy, sluggish but reliable **Cooper-BRM**. Lucien got a podium position from 3rd in Monaco, then scored 6th in the Belgian Grand Prix, but was less lucky as the season progressed. He did win that year's 24 Hours of Le Mans with **Pedro Rodriguez,** the crowning jewel of a long career. Unfortunately, he would die in 1969 testing an Alfa Romeo at Le Mans. Jules was the grandson of **Mauro Bianchi**, Lucien's brother who was also a driver. A **Ferrari** protege, Jules impressed greatly in the short period he drove a **Marussia** in Formula 1, being the first driver from one of the three teams introduced in 2010 to get points, two from 9th in the 2014 Monaco Grand Prix, and was also able to get the Marussia further up than anyone from that unhappy group ever did. Unfortunately, later in the season Jules crashed onto a service vehicle in the Japanese Grand Prix, sustaining serious injuries that would eventually lead to his

death in 2015, after a long hospitalization, the first F1 death since **Ayrton Senna**'s sad demise in 1994.
Lucien: 0 titles, 0 wins; Jules: 0 titles, 0 wins

FUTURE MOVERS AND SHAKERS

Starting in the 60s, until the 70s, many drivers began building and racing their own F1 cars, those being **Jack Brabham, Bruce McLaren, Dan Gurney, John Surtees, Emerson** and **Wilson Fittipaldi, Chris Amon and Arturo Merzario**. Some future team owners and even an engine builder took shots at driver stardom before deciding their talents could be better used behind the pit walls.

Arturo Merzario was the last driver to initiate a team and drive his own car. (Kurt Oblinger)

Colin Chapman

Colin Chapman would start fielding his **Lotus** cars in the 1958 Formula 1 season, but in 1956 was entered in the French Grand Prix as a driver by **Vanwall**. He did extremely well in qualifying, finishing with 5th best time, but a practice accident meant a DNS for the British driver. Chapman is responsible for introducing many innovations to F1, including monocoque chassis, full commercial sponsorship in Europe, wings on cars and wing cars, wedge shaped cars, ground effects. Very effective, his cars also had a reputation for fragility. He died young, at 54 years old, of a heart attack.

Bernie Ecclestone

Who would dream the used car salesman would one day be a competitive team owner and the most powerful man in the sport for ages? He had enough sense to give up on a driving career, but did attempt to qualify a self-entered **Connaught** B in the 1958 **Monaco** and British Grand Prix. In both cases Bernie did not qualify, and in Britain the car was driven by **Jack Fairman**. At any rate, the car was antiquated for 1958 standards, and never very fast on its heyday, so his attempt at Monaco is referred as "not serious" in most literature. After being **Rindt**'s manager, Bernie acquired **Brabham** in 1971 and revolutionized the financial side of the sport. He became a very wealthy man, but as he says to anyone who will hear, he made tons of people rich as well.

Roger Penske

Talk about wealthy people in racing, and the name of **Roger Penske** comes to mind. The American was actually quite a good driver in his younger years, and took part in the U.S. Grand Prix of 1961 and 1962, driving a **Cooper** and then a **Lotus**. He did not score any points, but finished both times, and then continued in the more sedate world of U.S. sports car racing. Penske fielded his own Formula 1 team between 1974 and 1976, and it won the Austrian Grand Prix in this last year. It is said that the death of his friend and collaborator **Mark Donohue** in the 1975 race took some of Roger's desire to stay in the category, but perhaps the idea was to build Penske's facilities in England, which eventually built Penske's very successful Formula Indy machines for the next couple of decades.

Penske was briefly in F-1 as a constructor, from 1974 to 1976. The cars were raced by others in the 1977 season. (Alejandro de Brito)

Brian Hart

Not a bad driver, Hart ran in a single Grand Prix, in the Formula 2 Section of the German Grand Prix of 1967. Driving a Ron Harris designed **Protos**, Brian was 12th on the road and fourth in class. He would not return. Later on Hart became well known for his engines, for many years preparing some of the best Formula 2 engines around. He took the big jump to Formula 1 in 1981, being one of the first builders of turbo engines initially provided to the Toleman team. The maiden season was discouraging, both **Henton** and **Warwick** could not get the cars to qualify. However, with time Warwick got some encouraging results, running closer to the front in 1983, and scoring nine points. The Hart engine was never quite a front runner in F1, but did get a pole in **Teo Fabi's** hands, in 1985, and equipped **RAM, Footwork, Jordan** and **Arrows** cars, besides Toleman.

The distinctive Toleman, with a Hart engine. (Kurt Oblinger)

Former GP drivers who ran Grand Prix operations as constructors after retirement include **Alain Prost, Aguri Suzuki, Alessandro de Tomaso, Jackie Oliver, Alan Rees, Guy Ligier** and **Gerard Larrousse**. Others, such as **Niki Lauda, Helmut Marko, Reg Parnell, Tim Parnell, Jackie Oliver, Bob Gerard, Vic Elford** and **Bobby Rahal**, had managerial/team ownership roles.

Guy Ligier became a F-1 constructor in 1976. Here is the 1979 car. (Alejandro de Brito)

INSUFFICIENT CHALLENGES

One can have unending discussions that Formula 1 has mostly been about domination by one team or driver over the other, since 1950. There is some element of truth to that, for even the 1975 season, in which 9 drivers driving for 6 constructors won races, was amply dominated by **Niki Lauda**. Generally, not always, the champion wins more races than others. However, there are different degrees of domination, even for a prevailing driver/constructor may actually face a serious challenge from a weaker opponent. But here are some cases in which the challenge was obviously insufficient.

TALBOT X ALFA ROMEO, 1950-1951

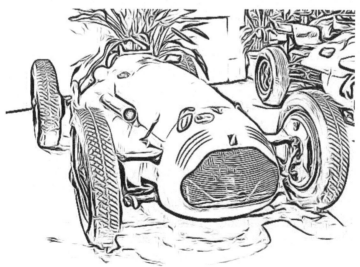

The French Talbot Lago was soundly beaten by Italian competition in 1950-51

Talbot-Lagos, or Lago-Talbots, equipped with 4.5 liter normally aspirated engines, were amply entered in the first two championships, in fact there were five of them against four **Alfa Romeos** at **Silverstone**, the first round of the 1950 championship. A pattern quickly emerged: although the Alfa Romeo was basically a 30s design, equipped with a supercharged 1.5-liter engine, it was much faster and powerful than the French cars: Alfa ended 1-2-3, while Talbots came in 4-5, two laps behind.

During the rest of season as many as seven Talbots were entered in World Championship events, and the best Talbots could do was a couple of 3rds by **Louis Rosier** in Switzerland and Belgium. By the end of the season Ferraris were also performing better than the Talbots, and things looked grim for 1951. The final score was Alfa 6 x Talbot 0.

At **Bremgarten,** in the first race of 1951, Talbots were again numerically superior, and seven of the cars were entered. Rosier was again the best of the lot, finishing 9th, but 3 laps behind the winning Alfa. The pattern continued in the rest of the championship: many Talbots on the grid (as many as eight at the Nürburgring), with all winning being done by Alfas and then, Ferrari. By the end of the season, Talbot's best result was a fourth by Rosier, in the Belgian Grand Prix. Score: Italian cars 7 x Talbot 0.

That was the last one saw of the French cars in the championship, which was contested by 2-litre Formula 2 cars in 1952-53, even though the name was restored in 1981, as the **Ligier** was rebadged Talbot Ligier. Talbot would finally feel the taste of victory, a couple of wins by **Laffite** in the second half of the season.

Credit must be given to Talbot's best performer Rosier. The Frenchman would win two editions of the non-championship Dutch Grand Prix, in 1950 and 1951, in the absence of Alfa Romeo. However, one of Talbot's feat will surely never be repeated. Rosier, always him, ran an altered Talbot grand prix car T26C (with fenders, spare tires, two seats) in the 1950 **24 Hours of Le Mans**, and won the race partnered by his son. Mind you, he did so driving 23 hours! So there was something to smile about, after all.

FERRARI WORLD, 1952-1953

Before you go into a rampage claiming that back in the old days there was tons of competition in Formula One and no domination by teams, consider this: in the two seasons the championship was run under Formula 2 regulations Ferrari won all but one F1 race and **Alberto Ascari** not only won both titles, but also won nine straight F1 races (plus two out of this sequence.).

It was not due for a lack of competitors, if you are wondering. Ferrari did not dominate the fields numerically, and quite a few other constructors raced in those two seasons: **Maserati, Cooper, HWM, ERA, Aston-Butterworth, Gordini, Veritas, AFM, EMW, Frazer Nash, Connaught, Alta, BMW, OSCA, Simca-Gordini**. With the possible exception of Maserati, none offered any resistance to Ferrari dominance, which finished most races 1-2, a few 1-2-3. **Fangio**'s absence was a factor only in 1952, when he was sidelined with injury. He raced in the entire 1953 season. Besides Ascari, **Piero Taruffi** won one race in 1952 (Ascari was off to Indianapolis), and **Hawthorn** and **Farina** won one apiece in 1953. Ferrari's only failure was at Indy, where the Italian lasted only 19 laps.

THE WORLD AGAINST MERCEDES, 2014-2020

There has never been a longer period of domination in Formula 1 than Mercedes' takeover of the World Championship in the 2014-2020 era, a whole of seven seasons. It was just not a matter of winning, it simply and utterly destroyed the competition during these seasons. The German manufacturer had quit as a works team at the end of the 1955 season, and returned in 2010, having bought the **Brawn** team. Things were not easy in 2010-2012, when Mercedes won a single race and not even the talents of **Michael Schumacher** could ensure the ultimate success of the team. However, with the arrival of **Lewis Hamilton** in 2013 the team became more competitive, winning three races.

Mercedes dominated the 2014 to 2020 seasons.

All of that changed when the new turbo 1.6-liter hybrid power plants were introduced, in 2014. Right away Mercedes dominated the field, winning 16 of the 19 races, and getting 18 poles. Lewis Hamilton and **Nico Rosberg** finished 1-2 in the championship, and Mercedes won the Constructors Cup.

Mercedes continued to dominate in the next six seasons, as Lewis won the 2015 then the 2017-2020 titles, as Nico Rosberg took the 2016 crown. During the period, a total of 138 races, Mercedes won 102 (Lewis: 73, Nico: 20 and **Valteri Bottas**: 9) and got 117 poles. The competition, **Ferrari, Red Bull, Racing Point** and **Alpha Tauri**, won a total of 36 races: effectively no one offered Mercedes a proper challenge until the 2021 season, when **Max**

Verstappen and Red Bull appeared to have turned the tables.

MCLAREN, 1988

Despite Mercedes domination in latter years, it palled in comparison to **McLaren-Honda's** performance of 1988. Nothing seemed able to stop the McLaren steam roller and the **Senna-Prost** intra team battle. It should be noted that this was the last season in which turbo engines were allowed until the current era, and most of the field had 3.5 liter normally aspirated engines. Boost was limited and fuel tank size reduced in the hopes of giving the normally aspirated engines a theoretical chance. Honda, which had partnered with **Williams** in the 1983-87 seasons, winning together the 1987 championship, shifted camps in 1988, following **Ayrton Senna** to McLaren. **Lotus** got to keep the Japanese engines, but except for some form shown by **Piquet** in the early season, seemed no match to McLaren. Other teams sporting turbo engines were **Arrows (Megatron,** a rebadged **BMW), Osella (**a rebadged **Alfa),** and **Zakspeed**. Then there was **Ferrari**. The Prancing Horse seemed in fact the only plausible alternative to McLaren's domination.

The Italian team did what it could. **Gerhard Berger** was the only non-McLaren to get a win and a pole position and three fastest laps. **Alboreto, Nannini** and **Mansell** also managed fastest laps. Senna got 13 poles, 8 wins to Prost's 7. In the end, Prost, who had 7 wins and 7 second places had more points than Senna, but had to drop some results as did Senna, losing the third title (90 x 87 in Senna's favor). The deciding factor was the rainy Japanese Grand Prix, where Senna brilliantly recovered from stalling on the grid.

WILLIAMS X MCLAREN, 1992

McLaren had won the titles from 1988 to 1991, and was still everybody's favorite for the 1992 title. **Senna** and **Honda** were still onboard, and **Gerhard Berger** could be counted on as a solid number two. However, **Williams-Renault** did give McLaren a harder time in 1991, than other challengers had in 1988 to 1990. They had **Mansell** who incredibly had not won a championship yet, and Williams would love to beat Honda, which left it without a competitive turbo engine in the end of the 1987. Vengeance is a dish one eats cold, as the Italians say.

Even so, no one could believe the level of domination Williams-Renault had in the early season. Mansell won the first five races from pole, and **Riccardo Patrese** finished 2nd in four of them. Senna, who only had a couple of 3rds so far, won at **Monaco**, followed by a Berger win in Canada. Mansell recovered his form and won the next three races (Patrese finishing 2nd in two of them), and would win another one, sealing the championship early in Hungary. Patrese would also win a race, and finish second in the championship. In the end, Williams-Renault won 10 races, Mansell would break the record for most wins in a season, with nine. McLaren would still win three more races, but the best Senna could do was fourth in the Championship, for even a pesky **Michael Schumacher** got more points than him. Williams got all pole positions, except at Canada, where Senna got the upper hand. In the end of the season Honda left Formula 1 one more time, while, surprisingly, Williams did not re-sign Mansell for the 1993 season, instead choosing Prost. It also gave Patrese the cold shoulder.

STATISTICS

The question is, who was the best of all-time? Subjectively, it is a matter of opinion. I am positive **Taki Inoue**'s mom thinks her son was the best. Objectively, stats can be very deceiving if you want to prove your favorite driver is the all-time top dog. At best, one can say **Clark** was the best at a certain period, **Fangio** in another, **Schumacher** another, etc.

The fact is, the parameters have changed so much since the 1950, that comparing one driver from the early 50s to a 2020s driver is unfair to both.

Consider this. The World Championship today has over 20 races, while many seasons in the 50s were a collection of 6 to 7 races. Thus Fangio, widely reckoned to be the best driver of the era, took part in only 51 races between 1950 to 1958 (he did not participate at all in 1952, and only ran two races in 1958). That is just a little over two seasons for a current driver. If you consider his stats, such as wins, poles, fastest laps and laps in the lead, he is proportionally unbeatable, compared to the current generation. A top driver today retires with much more than 200 GP participations: **Lewis Hamilton** has won tons of races, but at the time of writing was nearing the 300-GP mark, so his 103 wins do not stack favorably against Fangio's 24 from 51 races.

Even laps in the lead can be a deceiving stat. Many of the racetracks in the 50s to early 70s were very long indeed: **Nürburgring, Spa, Reims, Clermont-Ferrand, Interlagos**, and one-offs such as **Avus, Sebring, Casablanca** and **Pescara** all exceeded 7 km, while most current tracks are about 4 km long. A German GP from the period was 14 laps long, Nürburgring races as of late had almost 67 laps. When you change the stats to kilometers in the lead you get a fairer basis for comparison.

The most useless stat is points earned. A winner only got eight points in the 50s, and only the top 5 scored, while nowadays a winner takes 25 points home, and the 10th placed driver also takes a point. With the new qualifying races a weekend can be even more profitable. In other words, points serve for nothing in the wider scheme of things.

Current drivers, even the less successful ones, tend to stay much longer in F1. Longer commercial contracts, a driver's social media value, even appearance all weigh in the decision. Way back, you basically had one to three seasons to perform, or else you were shown the door real fast. One should also consider the fact that the sport was much more dangerous back then, and drivers often missed races or entire championships due to injury.

There are also different age groups. **Fangio** was 38, almost 39 when he started the **Silverstone** race in 1950, and many of the drivers on the field had raced before 1939 and **Villoresi** and **Etancelin** were past their 50s when they took part in the Championship. This, of course, was a result of World War II, but **Graham Hill** debuted in F1 when he was 28, **Jack Brabham**, 29. The latest crop of F1 driver debuts at 18, 20 years of age. **Max Verstappen** was not even 18 when he made his debut, and just above 18 when he won his first Grand Prix.

Different driving skills are also required in different eras. Races were 3 hours long in the early days of the world championship, tires skinny, brakes flimsy, surfaces of questionable quality. Add to that, you had to pay attention to your gear shifting, look at pit boards and feel the car. Some believe this makes a current driver's mission much easier, but it is not. Different skills are involved, and different problems arise. There are more G forces on the neck, drivers have to deal with strategists talking on their ears all the time (just ask **Räikkönen**), keep an eye on the readouts, be self-conscious that their every move is being recorded by an onboard camera – all these things are sources of stress. Plus the financial stakes are much higher now, by n-order of magnitude, and you have to be mindful of your every move on and off track, or else pay the price in social media the next day.

Should **Clark**'s accomplishments be dismissed, because he drove mostly down on power 1.5-liter cars with less power than a current F3 car, as opposed to the 1000 HP turbo hybrid engine beasts of today? Are the feats of drivers in the 3.0-liter Cosworth era, when most drivers drove very similar cars more noteworthy than those of a driver with a team of hundreds of engineers to provide proprietary upgrades with every other race?

Next time you get into a heated argument as to who was the all-time best, just smile. No one can win this argument.

THE U.S. GRAND PRIX SAGA

Formula 1 commercial rights have been bought by an American company, and the American footprint can be seen all over the sport now. No wonder that now there are two healthy U.S. Grand Prix, and there is a third on the way. As you will read below, three U.S. races in the same season were held in the past but it did not end well. It is unlikely that the same will happen now, for a huge effort (and money) is being put into making the sport more palatable to U.S. tastes.

Be that as it may, the relationship between the World Championship for Drivers and the U.S. has always been tense. Sometimes there were several races in the same season, sometimes no race at all. By far, the U.S. is the country with the largest place of venues used in the World Championship.

In the 1950s there was basically no interest in Formula 1 in the USA. Besides **Harry Schell**, who was just as French as he was American, drivers from this side of the pond did not race in F1. In fact, road racing was considered by American fans an underclass of racing. The compromise was adding the **Indy 500** as a round of the World Championship. Indy drivers would not race in F1, Europeans would not race at Indy, except for **Ward** at Sebring, 1959, and **Ascari** at Indy 1952. But at the end of the day, the names of American drivers who got the top 5 placings at Indy were on the final results from 1950 to 1960, so they have to be taken into consideration.

By the late 50s, a number of American drivers such as **Phil Hill**, **Dan Gurney** and **Carrol Shelby** had raced successfully in Europe, so the time was ripe for a proper round of the championship in the USA. That happened at **Sebring**, Florida, 1959, a race won by **Bruce McLaren**.

That season and the next the Indy 500 would still be a part of the World Championship, so there were two U.S. rounds of the Championship for a couple of seasons, 1959 and 1960.

As the Sebring venue did not quite work out, the 1960 U.S. Grand Prix was moved over to **Riverside**, California, a race won by **Stirling Moss** and the **Indy 500** was won by **Jim Rathmann**. That year was the last time the great race was included in the World Championship.

A more permanent home of the U.S. championship was finally found in 1961, **Watkins Glen**, a sleepy town in Northern New York state which became a stable site for a U.S. Grand Prix until 1980. The 1961 GP was won by **Ireland**, and **Dan Gurney** finished a great second, driving a works **Porsche**.

Formula 1 returned to California in 1976, as the fifth U.S. venue debuted. The street race at **Long Beach** was a popular stop in the calendar, being the only street race besides **Monaco** at the time. The layout of the track also allowed more overtaking than the streets of Montecarlo.

Long Beach was the first of many street courses used as U.S. Grand Prix venues. (Kurt Oblinger)

Watkins Glen was dropped from the calendar in 1981, as it became clear that the city and surrounding area had insufficient infrastructure to welcome such a world event. Races in similar places, such as **Anderstorp**, Sweden, had met the same fate. So instead, for 1981 a race was run in the Caesar's Palace parking lot track, in Las Vegas, while Long Beach remained firm.

The original **Las Vegas** race was not all that popular. Everything seemed very artificial and out of place, which is a pleonasm when one talks about Las Vegas. **Caesar's Palace** remained on the calendar for two seasons, lucky enough to be deciding races both times, but Las Vegas has not had good luck with car racing, and an Indycar race in the city ended in disaster in 2011.

Another venue was found at **Detroit**, 1982, so that particular season there were three U.S. races, none of them on racetracks. The idea behind the Detroit race was simple: symbolism. The city was considered the home of auto manufacturing in the USA, by the 80s more in tradition, than actuality. The US carmakers still had their headquarters there, but factories were mostly gone, plus the city needed an upgrade to its image, for it never quite recovered from 60s riots.

Long Beach would remain in F1 until 1983, jumping ship to CART the next year. For 1984, besides the Detroit Grand Prix, a new street race was added at **Dallas**. Poorly put together, under scorching heat, parts of the tarmac came apart and a large number of accidents sidelined a great part of the field. The fiasco was never repeated, so Detroit remained the only game in town for a while.

The "for a while" lasted until 1988. Detroit also moved over to the CART series.

A new venue was found for the 1989 season, Phoenix. In hindsight, holding a Formula 1 race in such a hot city, in June, was a mistake. So it was moved to a cooler month, March. It was not a long-term effort, and by 1992 there was no more Phoenix race. In fact, there was no U.S. race at all. For quite a while.

A U.S. Grand Prix would only be held in the U.S. again in 2000, of all places at Indianapolis, which built an infield and made changes to the track to welcome the F1 races. There is a bit of politicking here. A great tug of war was going on between CART and the IRL (connected to the Indy track through **Tony George**) for the future of Indy style racing. IRL promoted oval races, CART had races in ovals, street and road racing tracks. So IMS joined hands with the lesser enemy. On the good side, a U.S. F1 race was held at a permanent racing facility for the first time since the demise of the Watkins Glen race in 1980. The Indy F1 race was a popular one until the 2005 fiasco, when only six cars took the grid over safety concerns involving Michelin tires and the 2007 was the last race at Indy. Another hiatus ensued.

It was very difficult finding suitable racetracks in the U.S.A. to host a F1 event. Many of the good permanent tracks are located far from major metropolitan areas, and not willing to spend the millions required to bring them up to F1 standards. Indy fit the bill because it was a top facility. So the answer was building a track from scratch, which was achieved at the **Circuit of Americas**, in Austin, Texas.

The first Austin race took place in 2012, and except for 2020, due to Covid-19, races have been held in every season since then.

A new race was held in Miami Gardens, in 2022, dubbed the **Miami Grand Prix**, in the area surrounding a football stadium. The event was deemed a success, even though prices of artificially scarce tickets were outrageously high as most tickets were sold in the black market for higher rates than Monaco. For that you would get to see a couple of life size boats, parked on top of fake water and an artificial beach. The hype for the 2023 race has not been as high as the debut race, and ticket prices dropped considerably.

A third U.S. race looms in the horizon again, to be held at **Las Vegas** in late 2023. Whether the new Vegas event sticks this time out remains to be seen, and whether the U.S. market can support three F1 races is also open to discussion. The difference now is that Liberty Media will also organize the race, and will pull its weight behind it. Tickets for the race have sold for higher prices than even the 2022 Miami Grand Prix, and are more expensive than Monaco. We shall see.

In addition to these races, the non-race, the New York Grand Prix, deserves mention. The race in the most important city of the United States has been on the radar, on-and-off, for decades. Layouts in New Jersey, Wall Street area and Flushing Meadows, Queens have come to naught, as the organizers failed to navigate difficult New York politics. Long a dream of **Bernie Ecclestone**'s, the New York Grand Prix may still happen one day, as the city's reputation and situation has changed greatly in the wake of Covid-19 so the great city desperately needs to raise its profile once more.

MAJOR FAILURES

LOLA

It all started so well for Lola in F1 and ended the worst possible way. Lolas had been built for a few years, including Formula Junior/F2/F3 single seaters, and the company had a good reputation when the Bowmaker team ordered Formula 1 cars for the 1962 season. Driver **John Surtees** put a Lola on pole on debut in **Zandvoort**, but the effort ended in retirement. That was not the highpoint of the season, as Surtees did finish 2nd in **Aintree** and **Nürburgring**, and at the end of the season had 19 points. Lola had finished 4th in the Constructors Cup in its debut season.

The cars continued to be used by the **Parnell** team in 1963, by the likes of **Amon** and **Hailwood,** but no points were scored and Lola quietly faded away.

In the rest of the decade, Lola designed chassis for both **Honda** and **BMW**. The **Hondola**, as it became known, actually won at Monza, 1967. The BMW cars were used in the Formula 2 section of the German Grand Prix, referred as both **Lola-BMW** and BMW, and a F2 **Lola-Ford** non-started the 1967 event.

Lola would return to Grand Prix, sporting its own name, in 1974. The cars were fielded by **Graham Hill**, with sponsorship from Embassy cigarettes, and looked slightly cumbersome. The team scored a single point, and although Hill used the cars in initial races in 1975, he soon shifted to his proprietary **Hill** chassis.

The next Lola project would be the **Carl Haas** team, which fielded cars for **Alan Jones** and **Patrick Tambay** during the 1985 and 1986 seasons, equipped with **Ford turbo** engines. The effort did not bring the desired fruit, the sponsor was sold and appeared not interested in Formula 1, so Haas picked up his bags and left for greener CART pastures.

A sketch of the 1985-86 Haas Lola driven by Jones and Tambay.

Not all was lost, as Lola found a new client, **Gerard Larrousse**, who put together a team for **Philipe Alliot**, to race in the new **Jim Clark Cup** for normally aspirated engines in 1987. Not much could be expected, given the turbo competition had much more powerful engines, but Alliot got three 6th places during that season.

The cars continued to be run as Lolas in 1988 through 1991, first with Ford engines and then **Lamborghini** then back to Ford. However, entry lists were oversubscribed in this era and cars were required to pre-qualify.

Notwithstanding, Lola achieved reasonable success in 1990, scoring a total of 11 points as **Aguri Suzuki** finished 3rd in **Suzuka**, the marque's first podium since 1962!

In 1992 Lola was about to be out of the picture again, as Larrousse chose to buy the **Venturi** chassis, but found a new client for 1993, Scuderia-Italia, which ran a couple of **Ferrari** engined T93s for **Alboreto** and **Badoer**. .The cars were not successful and at the end of the season

Giuseppe Lucchini left F1 and left **Lola** without a client once more.

Lola was the top car manufacturer racing in the world at a time, and by 1977 had produced over 1000 units. Its cars had won a large number of races, surely numbered in thousands if one considers class wins, in all continents. However, the business had changed, and **Reynard** and **Dallara** were getting most contracts for single make series.

As a result for 1997 Lola decided to field its first proper works team. Plans were very optimistic and bold, including a proprietary engine in short order. Sponsorship from Mastercard theoretically would ensure the financial success of the team. A number of drivers were associated with drives, including **Jos Verstappen** and **Tom Kristensen**, but in the end the drivers were **Ricardo Rosset** and **Vincenzo Sospiri**.

The theoretical financial health of the team was based on a false premise. Instead of getting funds from Mastercard, the novel sponsorship deal involved convincing Mastercard holders to dish out between 50 to 1800 sterling to sponsor the team, in essence a *gofundme* proposition. Although both sponsor and team expected 0.32 percent of the 300 million cardholders would be compelled to dish out the money, the effort was a fiasco. Adding to that, the cars were simply not good, could not qualify under the 107% rule, so after the Australian and Brazilian races not only was the team done for, Lola was on the verge of liquidation. In the end, **Eric Broadley** lost control of his beloved company to **Martin Birrane**, and Lola has kept mostly a low profile since then. Not dead, but barely alive.

BAD GRAND PRIX ATTENDANCE

Figures for weekend GP attendance run in the hundreds of thousands in this day and age of TV, social media, credit cards and relative affluence. GP events are mostly welcomed annual international events in a world eager for entertainment at all costs, so Formula One prospers on TV and at the gate.

It has not always been so.

Anderstorp is very far from Stockholm and other major Swedish cities, in fact about 5 hours by car nowadays, and the Swedish Grand Prix had never been a success at the gate, despite of some surprising competition over the years. The 1978 race drew a crowd of only twenty thousand, a very discouraging figure, and another fact would render the event unsellable going forward. Both of Sweden's Formula 1 drivers, **Ronnie Peterson** and **Gunnar Nilsson**, would die in 1978, one from an accident at Monza, and the other from cancer. Thus the race was dropped from the calendar, never returning. Very few people saw the **Brabham** fan car in action.

Can it get worse? Yes, it can.

The 1967 French Grand Prix was a badly promoted affair. The race was normally held at **Rouen** or **Reims**, however, the event that year was to be held in the **Bugatti** circuit, at Le Mans, a couple of weeks after the 24 Hours. Depending on who you trust, between 8,000 to 20,000 bothered to attend and watch **Jack Brabham** win the race, with many other worthy drivers on track. The circuit was never used again, and Rouen was the 1968 venue, although France would use a number of other tracks in upcoming seasons (**Paul Ricard[Le Chatelet], Clermont-Ferrand** and **Dijon**) until settling on Paul Ricard for a while, and eventually Magny-Cours, and now back to Le Chatelet. Unbelievably, just a few years before 1967 the French race was known for paying the best prizes...

THE NEW TEAMS INTRODUCED IN 2010

In principle, it was not a bad idea. In the wake of the 2008 worldwide financial meltdown at least three manufacturer related F1 teams announced they would drop out from the sport: **Honda, BMW** and **Toyota**. Honda did so immediately, but the team morphed into Brawn and then Mercedes. As for BMW and Toyota, they would stay on for the 2009 season. Toyota vanished altogether, while BMW gave **Sauber** some room and time to find new supporters and investors to stay in business without corporate support.

Just so the grids would not fall beneath the 20-car comfort zone, the FIA called upon new applicants to join the world of F-1 without breaking the bank: in other words, from the start they would look more like privateer grid fillers from the 70s down, rather than the manufacturer based constructors they were actually replacing. No one trusted any of these teams would be front runners.

Among 15 applicants, four were chosen to join in the 2010 season, July 2019: **Campos, Virgin, Lotus** and **USF1**. The national diversity in team ownership became an obvious criteria for selection: the chosen group included Spanish, British, Malaysian and U.S. interests. Even a Serbian applicant appeared keen to join in the party, but its several moves were met with disdain.

Even before the start, one of the teams appeared to be in trouble. The USF1 team, an effort headed by **Peter Windsor**, did not appear to be a serious candidate, failing to come up with sponsors, proper facilities and designs. Strangely enough, the only chosen driver was an Argentine, **Jose Maria Lopez**, who lacked F1 experience. Soon the planned team disbanded and the Serbians tried to buy out the USF1 entry, to no avail. Nothing else was heard of USF1, and Peter Windsor lost quite a lot of credibility in the sport.

The team headed by **Adrian Campos**, a former Spanish driver changed hands even before competing, but remained in Spanish hands, and was renamed Hispania. The Malaysian team headed by **Tony Fernandes** had the best fan support, for it would bring back the Lotus name to Formula 1.

The three teams that showed up in Bahrain included two former GP winners. The cars all had the not-so-hot **Cosworth** CA2010 engine that also equipped **Williams** cars, and from the onset the status of grid fillers was apparent. Despite **Kovalainen**'s efforts in the Lotus, the best a new wave team could do in 2010 was 12th place, so no points for all three of them.

The pattern continued in 2011, even though Lotus got a fresh supply of **Renault** engines. In a weird turn of events, the engine supplier had acquired Lotus Cars, and argued in Court that the Malaysian team's use of the Lotus brand, acquired from **David Hunt** who bought the remains of the original Lotus operation in 1994, was illegal. **Proton**, which supported the team initially also sued, claiming breach. The legal wranglings lasted the entire year, and in the end, Renault prevailed. At any rate, Lotus was again the best of the lot, earning a 13th place during the season, while Hispania became **HRT**.

The news for 2012 were far from encouraging. Fernandes had bought **Caterham**, a company that produced Lotus 7 copies, and his team was renamed Caterham. Virgin's name was changed to **Marussia**, which had gained the support of Russian interests, and HRT was unsuccessfully put for sale at the end of the year. A couple of 13th places was the best the three teams could come up with.

Marussia was the only of the three new 2010 teams to scored points

2013 was an agonizingly cruel year for the teams in the back. In a season with few retirements overall, life became even harder for Caterham and Marussia, even though the latter had signed **Jules Bianchi**, who put in some brilliant performances in spite of the conditions posting a best of 13th in Malaysia.

In 2014 one of the grid filler teams finally got points, as Jules Bianchi, who looked set to a great future and did wonders in the Marussia equipped with a Ferrari engine, earned 2 points from ninth in **Monaco**. The good news was followed by bad as Jules was involved in a freak accident hitting a removal vehicle, on the wet in Japan, and after a very long hospital stay, became the first Formula 1 death since **Senna**'s demise in 1994. Caterham, on the other hand, barely survived the year. After skipping two races towards the end of the season, Caterham raced for the last time, as **Will Stevens** and **Kamui Kobayashi** competed in typical uncompetitive fashion.

Marussia was the sole remaining grid filler in 2015, and would last a couple more seasons. Still with Ferrari engines, its season best was a couple of 12th places by

Roberto Merhi and **Alex Rossi**. For 2016 Russian support was withdrawn, and the name once more changed to Pertamina Manor Racing, while cars went by the name of MRT, equipped with **Mercedes** engines. Pertamina, an Indonesian company, was involved to support countryman **Rio Haryanto**, but the love affair did not last the season. **Esteban Ocon**, then a Mercedes protégé did a few races to complete the line-up, while another Mercedes protégé, **Pascal Wehrlein**, got 1 point from a 10th place in Austria. In the same year Manor bowed out, while a new team entered Formula 1, doing a proper job. Whereas the grid-filler teams got a total of 3 points in seven years of racing, the American **Haas** team scored in 3 of the first 4 races of the season, including 5th place in Bahrain.

In the end, the 2010 teams fulfilled their main objective, which was to provide larger grids for fans. I suppose finding a die-hard fan for any of the three teams is very difficult, but they did provide a platform for some good drivers to finish off their F1 careers, such as Kovalainen, Trulli and Kobayashi, newcomers to get their feet wet in preparation for greater things, such as Ricciardo, Ocon and Wehrlein, and an opportunity for drivers unlikely to get chances anywhere else, such as Rossi, Lotterer and Haryanto. In essence, they were little more than embarrassments, quickly forgotten.

BAD FORMULA 1 ENGINES

Building a successful Grand Prix car is very difficult. Building a successful Grand Prix engine much more difficult, and even great competitors such as **Porsche** failed. The manufacturer's Formula 2 cars raced in a few Grand Prix during the 50s, and Porsche would become a works team during the 1.5-liter years. It did win, led on many occasions, but the company decided to focus on sports cars. It would return very successfully in the mid-80s, having built the **TAG** badged turbo engine that won 3 championships on a row, but disappeared after the 1987 season. Considering the TAG, one could expect the normally aspirated engine to be supplied to **Arrows** in 1991 would be a top power unit. The engine proved to be heavy, down on power and even unreliable, qualifying badly and retiring in every outing.

Subaru decided that sponsoring (branding) a GP engine program would be a good idea, and a tie-in was made with **Motori Moderni**, which had built turbo engines in the mid-80s. The flat 12 engine did not prequalify every time it was used by **Bertrand Gachot** in the 1990 season mated to a **Pacific** chassis, so the idea was scrapped after the agonizing season.

BRM was one of the top teams, constructors and engine builders in the early 60s, so when the new 3.0 liter regulations came into play, **Aubrey Woods** decided to revive an early 50s BRM concept that did not end well: a 16-cylinder engine. The complex engine in H configuration was used by both BRM and **Lotus** in the 1966 season, and only **Jim Clark** could extract enough speed from it to win a race, the US Grand Prix of 1966. For 1968 BRM decided to use a more conventional 12-cylinder engine, which eventually won 4 races in the early 70s.

As for unconventional engine configurations, **Life**'s W12 engine of 1990 must win the prize. The car was entered in most of the 1990 races, and did not ever come close to qualifying, despite **Bruno Giacomelli's** and **Gary Brabham's** efforts. The engine was designed by a former **Ferrari** engineer, **Franco Rocchi**, and mated to a company built chassis. The name seems almost comedic, for the whole outfit lacked life.

Bugatti was the winningest racing car constructor of the pre-war period, even though most of its great success came before the advent of **Mercedes** and **Auto Union** GP teams. It was a name to be respected, so there was some expectation that the Bugatti challenger that would debut in

1956 would be respectable, if not a world beater. The team got **Maurice Trintignant** to drive the mid-engined device in the race, and although 230 HP were claimed, the car qualified next to last and broke after 17 laps. It would never again be raced.

THE FOUR-WHEEL DRIVE CARS OF 1969

Getting all four wheels to drive a F-1 car was not a bad idea, except that it just made the cars heavier, and thus less competitive. The idea had been tried by Ferguson all the way back in 1961, but the car's only outing in the championship ended in disqualification. It did much better in non-championship races, although some of these could be very low-key affairs. A number of teams tried the concept in 1969, **Lotus, McLaren** and **Matra**. The Lotus 63 was entered for **John Miles** and **Mario Andretti**, by the works, and even by **Bonnier**, as a private entry. The car retired from most races, but did qualify close to the top ten on occasion.

The Lotus 63 – as well as other 1969 four wheel drive designs, was not successful

The M9A was the McLaren FWD car, and it was entered for **Derek Bell** in a single race, Britain, where it retired after staring 15th. The Matra MS84 was driven by both **Beltoise** and **Servoz-Gavin** in a few races, and although the latter finished 6th in Canada, the constructor decided against exploring the idea further. **Ferrari, BRM** and **Brabham** wisely did not try the concept.

TEAMS WHO WERE OUT OF THEIR DEPTH

Some Formula 1 teams were simply out of their depth, and perhaps should have never come to the racetrack, because they simply did not appear to be serious efforts at all. Some of them were products of the kit-car concept of the 1960s to 1980s, which made the most difficult part of a F1 racer widely available: a **Cosworth** engine mated to a **Hewland** gearbox.

Roberto Moreno drove for a number of small teams. At AGS at least he managed a point, in 1987.

A few names spring to mind, such as the **Andrea Moda** of 1992, a team fielded by **Andrea Sasseti**, built upon **Coloni**'s ashes. The team qualified for a single race, due to **Roberto Moreno**'s efforts but never looked race worthy elsewhere. In the overcrowded fields of the late 80s, early 90s, two other clueless teams were Coloni and **Euro-Brun**. Quite amazingly, Brazilian Robert Moreno was hired to drive for these three teams, and also **Forti** and **AGS**, two other basket cases. After the 1995 Forti fiasco, Moreno decided to concentrate on a career at CART.

The early **Williams** cars, the **Politoys** and **Iso-Marlboros** from 1972 to 1974, were not good cars and took a lot of talent and a load of strong will and a measure of courage to get the cars to go fast. Getting a hold of one of these cars three to four years later, by which time they were dated designs, and modify them to go Grand Prix racing and expect good results seemed an irrational idea. However this was done not by a single overly optimistic individual, but two.

The Apollon was just one of dozens of failed Cosworth powered projects of the 70s.

Loris Kessel had raced occasionally in the 1976 F1 Championship driving one of the **RAM Brabhams**, so for 1977 he tried to go one step further: fielding his own car. He took an old **Williams**, a FW03, tweaked it here and there, and voilà, called the car **Apollon**. The device was entered by the Jolly Club of Switzerland in the Italian Grand Prix, and needless to say, did not make the cut: 33rd out of 34 hopefuls.

Australian **Brian Maguire** had a FW04, a better car, which he entered as a Williams in the British Shellsport series, and managed to win a race. He also tried to enter the British Grand Prix in 1976 and 1977. In the first year, he was first reserve and did not compete. In 1977, he modified the Williams beyond recognition, and proudly called it the **McGuire** BM1. Needless to say, he did not make the cut, scored the 35th time out of 36. A while later he crashed it in Brand Hatch and died from injuries.

A number of other non-performing teams are noteworthy and will be scrutinized in later volumes, including **Connew, Maki, Shannon, Pacific, Amon, Merzario, Scarab, etc.**

Connew, one of three new makes that debuted in 1972. It did not last long.

Even very successful teams in other disciplines can be out of place in Formula 1. **Willy Kauhsen** had been a major force in Interserie and endurance racing, driving and running **Porsches**, and in 1975 ran **Alfa Romeos** in the World Championship for Makes, winning the title for the Italians. He then took the plunge into Formula 2 in 1976, running March-Harts, but in spite of having good drivers, **Klaus Ludwig** and **Ingo Hoffmann**, the team did not do well. He did what seemed to be sensible for 1977, buying the 1976 F2 championship winning **Elf-Renaults**, and renaming the cars **Kauhsen**. In this new guise the cars were very unsuccessful, so Willy sold them in Japan and South Africa and decided his place was Formula 1. Initially there were rumors that Willy had reached an agreement with **Kojima** and would race the Japanese cars in F1, which would not be a bad idea. In the end, he decided to build his own cars. The Kauhsen WK driven by **Gianfranco Brancatelli** was only entered in two Grand Prix, in Spain and Belgium, 1979, and did not do well, finishing with the slowest time on both occasions. Distraught, Kauhsen decided enough was enough, sold equipment to **Merzario** and disappeared from racing circles.

The **Arzani-Volpini** had a single attempt in the Formula One Championship, in the 1955 Italian Grand Prix. The car was no more than a revised old **Maserati** 4CLT, and **Luigi Piotti** did not start the race. A few months before the slow car was entered in the non-championship **Pau** Grand Prix, and driver **Mario Alborghetti** crashed it, dying from injuries. The constructor decided to go back to building lower-formulae cars and forget about the big time.

Bellasi was another Italian constructor that built cars for lower formulae, and won quite a few races at that level. Swiss **Silvio Moser**, a privateer who did score points with **Brabham**s in 1968 and 1969, decided to commission a **Bellasi** for Formula 1 racing, which was not a good idea. The Bellasi was entered in five races in 1970, and one in

1971, and only started in two, retiring both times. The constructor continued to do well in the Italian small engine single seater world.

The Rebaque HR100 was an afterthought.

Hector Rebaque had been driving F-1 cars for a while, including older **Lotus** cars for his team Rebaque, when a decision was made by the family to commission a proprietary car. In fairness to the designer and the team, there was not enough time to assesses whether the **Rebaque** was a dud or a car with potential. It was entered in three events, and qualified in one (Canada, 1979) 22nd out of 29th. Then the HR100 was shelved, as Rebaque became a Brabham driver in 1980-81, acquainting himself reasonably in the latter year. Rebaque would eventually win a CART race in 1982.

GREAT ACCOMPLISHMENTS

Certain Grand Prix facts were so massively outstanding that they can only be identified as great accomplishments. These are the great performances that are unlikely to be repeated and are worthy knowing about.

THE RHODESIAN ROULETTE OR WHOLE LOTTA LOVE

The South African Grand Prix was included in the World Championship for the first time in 1962, and from the onset until 1976 it was known for a peculiarity: several cars were entered by local teams for regional drivers. That is because there was a South African Championship often referred as South African Formula 1 Championship, that was contested by current regulation Formula 1 cars. Calling it a F1 Championship was a bit of exaggeration, as you will see in other parts of this book, but even locally produced cars like the **LDS**, the **Alfa Special** and other 1.5 Alfa Romeo Giulietta engined cars that did not appear elsewhere would pop out in the race.

The 1967 event was already different from the others held so far, in that it was being run in **Kyalami**, a racetrack about 5000 ft above sea level, where all further editions of the race were held. In qualifying, the 5th place driver drew some interest. Rhodesian **John Love**, who had driven in Formula Junior in Europe, managed to put his 2.7 **Climax** engined **Cooper** on the second row, ahead of many prominent drivers, including former world champions.

This was no fluke. Love drove in the leading group all day, despite the down on power engine, but the race seemed to go **Denis Hulme**'s way, who would always do well at Kyalami in the future. The New Zealander was 1 minute ahead, and then went into the pits as his Brabham-Repco developed problems. At the time second place was being held by Love, who had a clean run so far and became the leader. The Rhodesian continued to lead **Pedro Rodriguez**, on a works **Cooper-Maserati**, by as much as 24 seconds, but his luck ran out: with seven laps to go, he pitted for fuel. He lost the lead, disappointing the local crowd, but still managed to hold on to second. This turned out to be the best performance by a locally entered driver in South Africa by a great margin.

THE BEGINNING OF A REVOLUTION

The Argentine Grand Prix was held for the first time in 1953, and except for that first race, which had a team of **Cooper-Bristols** on the grid, it was simply overlooked by British teams, until 1958. Transporting the cars all the way to Argentina was very expensive, but Italian teams did not mind, as they usually managed to sell old cars, parts and engines to local and regional drivers. The British cars were not used in South America at the time, so there was no additional financial compensation for them. The French **Gordini** and German **Mercedes** also made the trek, on occasion.

Stirling Moss beat all 2.5 liter cars with his Climax engined Cooper in Buenos Aires, 1958 (Alejandro de Brito)

By 1958 the number of Italian cars in Formula 1 was decreasing, as the **Maserati 250 F** was getting dated, and Ferrari was no longer fielding five car teams, so the outlook for the Argentine Grand Prix was not bright. To make up the numbers, 3-time world championship runner up **Stirling Moss** was entered by **Rob Walker** on a 1.9 **Cooper Climax** with the engine on the back. Moss was a **Vanwall** driver at the time, but his team decided to save the money that would be spent in the long Southern adventure, releasing the driver. Moss qualified the small car 7th out of ten.

Rear engined GP cars were not novelty. **Auto Unions** had ran from 1934 to 1939 as rear engine designs, and the **Bugatti** 1956 car was mid-engined, although it raced only once and had a forgetful performance. The Coopers of 1953 were front engined, but Cooper had used the rear engine T43 in 1957, which, although not a front runner, did not disgrace itself. Even the second F1 race of 1950 had a rear engine Cooper on the grid.

That day Moss had to fight three **Ferraris** and six **Maseratis,** all of which had more powerful 2.5 liter engines than him. On pole was **Fangio,** followed by future champion **Hawthorn** and **Collins**. The latter named driver retired right on the first lap, with driveshaft issues. **Behra,** Fangio and Hawthorn appeared more able to take command of the race, but right on the early stages Moss showed the little Cooper should not be underestimated. He eventually passed Fangio for the lead, as the Maserati tires gave up the ghost. Moss continued in the lead to the end, as tires on the heavier Italian cars went to shreds. The Ferrari team figured Moss would need to stop as well, and in fact, the Cooper's tires were done for at the end of the race, but Moss kept the car on the road, and won the race by a little over two seconds. That was no isolated lucky victory, as **Trintignant** would win the next round, Monaco, in the same car, beating Moss' Vanwall, among others. The future had arrived.

A POINT PROVEN

By 1996, a pattern had emerged in the **Williams** Formula 1 team. The team became known for winning championships one year, only for the champion to leave the next season. It first started in 1987, when **Nelson Piquet** won the crown, but as **Honda** decided to shift its engine supply from **Williams** to **McLaren** and Nelson wanted to keep running the Japanese powerplants, the Brazilian went to **Lotus**. Then, from 1992 to 1993, **Nigel Mansell** finally won the championship, but team and driver could not agree on a proper retainer, thus no Nigel in 1993. The English driver went to CART, winning the championship sensationally. Nigel's place was taken by **Alain Prost**, who won the championship, then retired at the end of the season. I suppose the next Williams champion could see the writing on the wall: **Damon Hill** won the 1996 title but was not re-signed by Williams, apparently there was a disagreement on retainer. There were no top rides available, so Damon ended up driving the **Arrows-Yamaha** in the 1997 season, a less than successful combination. Arrows had been around since 1978, and had never won a GP. Yamaha's engines had equipped many a team since 1989, and had never won a race, nor looked competitive. Damon's season looked quite a disaster, and a single point from 6th in Britain had been scored by the time the circus came to the **Hungagoring**. Anyone that thought Damon Hill was a second rate driver had to reconsider such belief that day. Damon qualified the car 3rd, 0.372 seconds slower than **Michael Schumacher**, and was competitive right at the start, overtaking championship leader **Villeneuve**. He then caught up with Michael, and simply passed him on the 11th lap, keeping on leading the rest of the race. It did not go his way, a hydraulic failure caused by an inexpensive part gave Jacques Villeneuve a chance he did not really deserve, and Hill was passed halfway through the last lap. Damon mustered enough momentum to finish 2nd and prove he was world class.

SHARED DRIVES

Nowadays a shared drive in a Formula 1 race would appear odd. After all races are relatively short, even though in other categories even 1-hour races involve a change of driver. However, Grand Prix racing historically had longer races than the current hour and a half affairs, in fact, in the 20s they lasted 10 hours in 1931.

These 10-hour GP races did not last long, yet, in the early years of the World Championship races generally lasted about 3 hours, the Indy 500 a bit more, almost four hours, so relief sometimes was required and advisable.

Drivers were never entered in pairs in the world championship, do not be mistaken. The reasons for taking over a car were the physical conditions of a driver, clearly an issue in the 1953 and 1954 Indy 500s and the 1955 Argentine Grand Prix, or the fact that a team leader's car broke. Until 1957 drivers of a shared car were able to share points, which came in handy, especially in **Fangio**'s 1956 campaign. Two of Fangio's wins were achieved in shared drives (France, 1951, with **Fagioli**, and Argentine, 1956, with **Musso**) and **Moss** also benefited from a shared drive win (Britain, 1957, with **Brooks**).

So here are the shared drives in the Championship, from 1950 to 1960, when they were last allowed, with some comments where suitable.

1950
Britain, Fry/Shawe Taylor, Maserati, 10th place
Indy, Banks/Agabashian, Maserati-Offy, retired
France, Etancelin/Chaboud, Talbot, 5th
Italy, Taruffi/Fangio, Alfa Romeo, retired
       Serafini/Ascari, Ferrari, 2nd

1951
Indy, McGrath/Ayulo, Kurtis-Kraft, 3rd
France, Fagioli/Fangio, Alfa Romeo, 1st
    Fangio/Fagioli, Alfa Romeo, 11st
    Gonzalez/Ascari, Ferrari, 2nd
Italy, Bonetto/Farina, Alfa Romeo, 3rd

1952
Switzerland, Simon/Farina, Ferrari, retired
France, Fischer/Hirt, Ferrari, 11th
    De Graffenried/Schell, Maserati, retired
Holland, Landi/Flinterman, Maserati, 9th

1953
Argentina, Trintignant/Schell, Gordini, 7th
Indy 500 – This was the Indy 500 from hell. Temperatures remained in the upper 90s (36°C) throughout the end, and track temperatures exceeded 130° F (54°C). As a result, many drivers required help from colleagues, some of them drivers who did not qualify or were just sitting about on race day. Winner Vukovich did not enlist anyone's help, but in some cases, as many as three drivers drove the cars. One driver, Carl Scarborough, was taken to the infield hospital and died from heat exhaustion!
    Hanks/Carter, Kurtis-Kraft, 3rd
    Agabashian/Russo/ Kurtis-Kraft, 4th
    Rathmann/Johnson, Kurtis-Kraft, 7th
    Bettenhausen/Stevenson/Hartley, Kuzma, 9th
    Scarborough/Scott, Kurtis-Kraft, 12th
    Holland/Rathmann, Kurtis-Kraft, 15th
    Ward /Linden/Dinsmore, Kurtis-Kraft, 16th
    Faulkner/Mantz, Kurtis-Kraft, 17th
    Webb/Thomson/Holmes, Kurtis-Kraft, 19th
    Hoyt/Stevenson/Linden, Kurtis-Kraft, 23rd
Belgium, Claes/Fangio, Maserati, 3rd on road, unclassified
Germany, Villoresi/Ascari, Ferrari, 15th
Switzerland, Fangio/Bonetto, Maserati, 4th
    Bonetto/Fangio, Maserati, retired
Italy, Mantovani/Musso, Maserati, 8th

1954

Indy 500. The 1954 Indy 500 was just as hot as the 1953 race, but perhaps, less shocking. Vukovich again won the race without relief, a testimony to his obvious endurance and stamina. This time around no one died from the heat, and as many as five drivers took stints in individual cars.

    Ruttman/Carter, Kurtis-Kraft, 4th
    Russo/Hoyt, Kurtis-Kraft, 8th
    Cross/Persons/Hanks/Linden/Davies, Kurtis-Kraft, 11th
    Stevenson/Faulkner, Kuzma, 12th
    Carter/Teague/Jackson/Bettenhausen, Kurtis-Kraft, 15th
    Elisian/Scott, Stevens, 18th
    Armi/Finder, Kurtis-Kraft, 19th
    Hanks/Davies/Rathmann, Kurtis-Kraft, retired
    Ward/Johnson, Pawl, retired
    Hartley/Teague, Kurtis-Kraft, retired
    Thomson/Linde/Daywalt, Nichels, retired
    Linden/Scott, Schroeder, retired
    Rathmann/Flaherty, Kurtis-Kraft, retired
    Webb/Kladis, Brommel, retired
    Duncan/Fonder, Schroeder, retired

Belgium, Hawthorn/Gonzalez, Ferrari, 4th
Britain, Bira/Flockhart, Maserati, 19th
    Villoresi/Ascari, Maserati, retired
Germany, Gonzalez/Hawthorn, Ferrari, 5th
Italy, Maglioli/Gonzalez, Ferrari, 3rd

1955

Argentina – January in Argentina can be very hot, in fact, that is the height of the summer. The 1955 Grand Prix was held in grueling conditions, so many drivers, including front runners, required relief. A few cars had three drivers. Although Fangio was a man in his early 40s, he did not require relief, and drove solo, not missing a beat and winning by over a minute

    Gonzalez/Trintignant/Farina, Ferrari, 2nd

Farina/Maglioli/Trintignant, Ferrari, 3rd
Hermann/Kling/Moss, Mercedes, 4th
Schell/Behra, Maserati, 6th
Musso/Mantovani/Schell, Maserati, 7th
Bucci/Schell/Menditeguy, Maserati, retired
Mantovani/Musso/Behra, Maserati, retired
Castellotti/Villoresi, Lancia, retired
Monaco, Behra/Perdisa, Maserati, 3rd
Perdisa/Behra, Maserati, 8th
Taruffi/Frere, Ferrari, 9th
Indy, Bettenhausen/Russo, Kurtis-Kraft 2nd
Faulkner/Hometer, Kurtis-Kraft 5th
Belgium, Mieres/Behra, Maserati 5th
Britain, Hawthorn/Castelotti, Ferrari, 6th
Wharton/Schell, Vanwall, 9th
Rolt/P.Walker, Connaught, retired

1956

In this season, some of Fangio's shared drives helped him get enough points to prevail. This was by far the Argentine's toughest championship win, and one cannot say he enjoyed his time at Ferrari. His challenger Moss also enlisted the help of teammate Perdisa on occasion, taking over the Italian's car more than once,
Argentina, Musso/Fangio, Lancia-Ferrari, 1st
Landi/Gerini, Maserati, 4th
Uria/O. Gonzalez, Maserati, 6th
Monaco, Collins/Fangio, Lancia-Ferrari, 2nd
Fangio/Castellotti, Lancia-Ferrari, 4th
Bayol/Pilette, Gordini, 6th
Indy, Elisian/E.Russo, Kurtis-Kraft, retired
Belgium, Perdisa/Moss, Maserati, 3rd
France, Perdisa/Moss, Maserati, 5th
Hawthorn/Schell, Vanwall, 10th
Britain, De Portago/Collins, Lancia-Ferrari, 2nd
Castellotti/De Portago, Lancia-Ferrari, 11th
Germany, De Portago/Collins, Lancia-Ferrari, retired
Musso/Castellotti, Lancia-Ferrari, retired

Italy, Collins/Fangio, Lancia-Ferrari, 2nd
   Fangio/Castelotti, Lancia-Ferrari, 8th
   Maglioli/Behra, Maserati, 8th
   Villoresi/Bonnier, Maserati, retired

1957
Argentina, Gonzalez/De Portago, Lancia-Ferrari, 5th
   Perdisa/Collins/Von Trips, Lancia-Ferrari, 6th
Monaco, Von Trips/Hawthorn, Lancia-Ferrari, retired
   Scarlatti/Schell, Maserati, retired
France, McDowell/Brabham, Cooper, retired
Britain, Brooks/Moss, Vanwall, 1st
   Trintignant/Collins, Lancia-Ferrari, 4th
   Moss/Brooks, Vanwall, retired
Italy, Scarlatti/Schell, Maserati, 5th
   Simon/Volonterio, Maserati, 11th

1958
France, Lewis-Evans/Brooks, Vanwall, retired
Italy, Gregory/Shelby, Maserati, 4th (no points earned)

1960
Argentina, Trintignant/Moss, Cooper, 3rd (no points earned)

SCARY STUFF

Not all has been glamour, fun and games in Formula 1. From the start in 1950 until the early 80s many drivers lost their lives in accidents, and big shunts continued to this day, witness **Zhuo Ghanyu**'s recent accident in the 2022 British Grand Prix, which could have resulted in a death were it not for the driver head protecting halo. Although many old-timers criticize the sport's current overly cautions approach, with plenty of red flags and safety cars, the last F1 death, **Jules Bianchi's** in the wake of an accident in Japan, could have been avoided if more caution were exercised. Here are some of the moments I wish had never occurred.

In **Alberto Ascari's** case, the outcome of his Monaco 1955 accident was not fatal. Racing his **Lancia** in second place, behind **Moss**, the Italian may have become distracted over the huge smoke cloud left by the broken **Mercedes** engine. He then lost control of his car, flipped over the barrier and his car dove right into the water. Alberto was able to swim to safety, but four days later, testing a **Ferrari** at **Monza,** the twice world champion met death prompting the disbanding of the Lancia team.

The day October 6 would become known as a day of infamy for **Watkins Glen** habitues. For two years running, drivers died on that day in the New York state track. In 1973, **Francois Cevert**, tipped to become Tyrrell's number one driver in 1974, lost control of his car in practice, and his body was cut in half, dying instantly. That meant the sad withdrawal of the **Tyrrell** team, whose driver's **Jackie Stewart** had won the championship a couple of races back. In the next year October 6 was the race day. Austrian **Helmut Koinigg**, who had his first start in Canada, was driving a **Surtees** that day, which suffered suspension failure. The car did not go off too fast, but the lose armco, there to protect the driver, actually decapitated the Austrian as he crashed into it.

Mark Donohue's accident in Austria, 1975, was scary for a different reason. The American had a crash in practice, managed to leave the car, sat down near the stricken vehicle consciously, and all seemed fine. However, his brain injuries were profoundly serious, and Donohue died in the hospital. In an eerie turn of events, in the next season Donohue's team, **Penske**, had its single victory in the very same track that claimed its most famous son.

Mark Donohue's death in Austria, 1975 was rather unusual (Rob Neuzel)

Ferrari's 1982 was full of ups and downs and drama, atypical even for the Italian team, known for loads of both.

The red cars were some of the fastest in the early part of the season, and **Villeneuve-Pironi** seemed in for a treat. Most of the British teams boycotted the San Marino race at Imola, leaving the way open for an easy Ferrari win. Pironi won, but Villeneuve felt betrayed by his teammate. After the **Renaults** retired, Ferrari drivers were told to slowdown, which the Canadian did. The Frenchman continued business as usual, winning the race and Villeneuve's hatred, for the Canadian promised never to speak to him again. In the next race at **Zolder**, Belgium, two weeks after this event, under heavy rain, Villeneuve would die attempting to pass the slower March of **Mass** during practice. Both cars crashed and a sublime talent was gone. In sign of respect, Ferrari withdrew from that race, but Pironi continued to be the most likely contender, and by the German Grand Prix had amassed 39 points.

Gilles Villeneuve remains a very popular F1 figure forty years after his demise (Kurt Oblinger)

The Frenchman got pole position, but had a heavy crash in practice, and was sidelined for the rest of the season. His new teammate **Tambay** did win the race on Sunday, easing the pressure on Pironi. Further Ferrari efforts were not sufficient to ensure the driver's title to their hero. The team insisted the seat was Pironi's once he was fit to resume his car racing career, but that never happened. In 1986, after testing for a couple of teams (**AGS** and **Ligier**), having recovered full use of his legs, Pironi had to reconsider due to insurance issues: if he went back to GP racing he would need to repay back the insurance proceeds. So, instead, Pironi decided to pursue powerboat racing in 1987, and in an open sea event, in England, his boat tipped over killing Pironi and his crew.

The events in **Imola**, 1994, have been widely chronicled. Not only was **Senna's** accident shocking, but the fact that two drivers died during the weekend, while a third, **Barrichello** had a very nasty shunt, seemed totally out of place in current F-1. Multiple driver deaths in a same racing event were not uncommon: a 1967 Formula 3 race in **Caserta**, Italy, resulted in the death of former GP driver **Geki Russo** and others, another F-3, at **Rouen**, 1970, also had multiple deaths and a close-call, while three top drivers were killed in different accidents in the 1933 Italian Grand Prix. But that seemed a thing of the past. Many have tried to connect the Senna incident to the death of **Roland Ratzemberger** the day before, but Senna's accident was due to a car fault, rather than loss of control and concentration due to emotional reasons. Clearly there was no connection, they were freaky sets of circumstances.

LENGTHY RACING CAREER SPANS

Some drivers managed to stay around in Formula 1 for quite a long time, even though some listed here did just a few races, while others ran hundreds of times. So let us start with the unusual Lengthy racing career spans.

PETE LOVELY, 1959 to 1971

The American was a sports car champion in the SCCA, who decided to race in Europe in 1958. He joined Lotus, and was given his first chance in the 1959 **Monaco** Grand Prix, failing to qualify his **Lotus Climax**. His first GP startwas at **Riverside**, the 1960 US Grand Prix, where he finished 11th. He vanished for a while, returning in 1969, and almost got a point in the Canadian Grand Prix, again driving a Lotus. He also raced in the US and Mexican Grand Prix that season. In 1970 he actually tried to qualify his **Lotus 49B** in a few European rounds, failing in two but actually starting the British Grand Prix. His last two races took place in 1971, when he raced a Lotus 49/69 hybrid in the North American races, failing to get enough laps to be classified both times.

JAN LAMMERS, 1979 to 1992

Dutch driver Lammers started to race at a very young age in **Zandvoort,** and by 1978 had won the European Formula 3 championship. He looked set to a great future, as he was hired by **Shadow** for the 1979 season. Things did not work out much and his best finish was 9th. He impressed sufficiently to be hired by **ATS** for the first part of 1979, and surprised even the keenest follower of F1 by placing the non-competitive car fourth on the grid in **Long Beach**. The ride did not last long, and he did the rest of the season driving for **Ensign**. Then he returned to ATS for four races, got several chances in the **Theodore,** but only started in his home race, retiring. Then Lammers vanished from the F1 world. He became a top sports car driver, won **Le Mans**, did Indycars, and then, out of nowhere, appeared in a **March-Ilmor** for the Japanese and Australian Grand Prix of 1992. That, apparently, is the end of his F1 career...

ANDRE PILETTE, 1951 to 1964

Belgian Andre Pilette got some good results outside of F1, which ensured his name being remembered for rides, specially near the Belgian Grand Prix. He had a total of 9 starts, the first on a 4.5 **Talbot-Lago** T26C, in 1951, the last driving a 1.5 Scirocco-Climax in 1964. In between he raced **Connaught, Gordini, Emeryson, Lotus** and even a works **Lancia-Ferrari**. He got 2 points from fifth place in the 1954 Grand Prix.

HANS HERRMANN, 1953 to 1969

German driver Herrmann had an interesting Grand Prix career, that began in the 2-liter regulation year of 1953, passed through the 2.5-liter years, then the 1.5-liter era, ending in the 3.0-liter era. He first drove a **Veritas** in the German Grand Prix of 1953, finishing 9th. His most successful period was driving for **Mercedes** during the 1954 to 1955 seasons, a total of six races. His best result was 3rd in the Swiss GP of 1954, but he also got two 4th places, however a nasty **Monaco** crash sidelined him from racing the rest of 1955. From that point forward Herrmann raced little in Formula One, instead becoming a sports car specialist. He had a few outings driving Centro Sud and Bonnier **Maseratis** in 1957 to 1958, then drove a **Cooper** and **BRM** in 1959. He reappeared as a **Porsche** works driver in the 1960 and 1961 seasons, finishing 6th in the 1960 Italian Grand Prix. He also drove **De Beaufort's** car in the 1961 Dutch Grand Prix. His latter appearances occurred in the Formula 2 section of the German Grand Prix driving for **Roy Winkelmann**. He finished 11 in a

Brabham in 1966, and did not start the 1969 race, in which he was entered in a **Lotus** in respect for the death of his friend and teammate **Gerhardt Mitter**. Herrmann won in class in the **24 Hours of Le Mans** many times, always driving Porsches, and was also on the crew of the Porsche 917 that won the first overall victory for Porsche, in 1970, sharing a car with **Richard Attwood**. Having achieved this victory, Herrmann retired, and at the time of writing is still alive, in his 90s.

HARRY SCHELL, 1950 to 1960

Harry held both American and French citizenships, and was born in France. That obviously made things easier for him to become the first regular American GP driver. And he stuck around for quite a long period, driving in 56 races and scoring 30 points. He drove a vast range of machines, starting with a 1.1 **Cooper-JAP** in the **Monaco** 1950 race, then **Talbot, Maserati, Gordini, Ferrari, Vanwall, BRM** and ending where he started, with a Cooper. His best championship result was 2nd in the 1958 Dutch Grand Prix.

ANDREA DE CESARIS, 1980 to 1994

The Italian was widely criticized in the early part of his career, for many felt his connections with sponsor Marlboro were the only reason for his being considered for rides, even though he won many Formula 3 races in 1979 and a Formula 2 race in 1980. The fact is that Andrea had a very long, if not very successful career in Formula 1, evolving from a crash-prone youngster of the 80s to a dependable performer for midfield teams in the 90s. He did a couple of races for **Alfa Romeo** in 1980, then had a miserable season driving for **McLaren** in 1981, earning a single point from 6th in San Marino. He returned to Alfa Romeo for the 1982 and 1983 seasons and had pole and led at **Long Beach** and very nearly won the **Monaco** Grand Prix in 1982. He retired from most races in 1983, but finished 2nd in Germany. Andrea was hired by **Ligier** for the 1984 and

1985 seasons and ended in the points 3 times in the two seasons. The 1986 and 1987 seasons, at **Minardi** and then **Brabham** were full of retirements. In fact De Cesaris finished a single race in those two seasons, but at least was on the podium in Belgium, 1987. He was hired by one of the new teams, **Rial**, for the 1988 season and surprised most with a 4th place at **Detroit,** plus had some impressive qualifying efforts. In two seasons for Scuderia Italia, which ran **Dallara-Fords**, Andrea had the last podium of his career, in Canada, 1989, and again did wonders in qualifying for some races. In 1991 Andrea appeared a changed man and scored points in four races at new team **Jordan**, finishing nine of the last twelve races of the season. That was enough to convince **Tyrrell** to hire the Italian, where he scored points in four races with the Ilmor engined car in 1992, failing to score with the **Yamaha** powered car in 1993. In 1994 the end appeared near, but the Italian was called to replaced injured drivers **Barrichello** at **Jordan,** and then **Wendlinger** at **Sauber**, and scored points for two different manufacturers in his last season. He retired at the end of the season.

De Cesaris' season at McLaren was not successful (Kurt Oblinger)

MAURICE TRINTIGNANT, 1950 to 1964

Maurice, known as Le Petoulet, was already racing before the war, and although he was already 33 in 1950, remained active for many years. He raced a large number of machines in the World Championship: **Simca-Gordini, Ferrari, Gordini, Vanwall, Lancia-Ferrari, Cooper, Maserati, BRM, Aston Martin, Lotus, Lola** and the disastrous rear engined **Bugatti**. He is the first on this list to have won a Grand Prix, on both cases, the **Monaco** Grand Prix, in 1955 and 1958. He also won many races outside of Formula 1 and to his credit, scored 2 points in his last season, 1964, from fifth in the **Nürburgring**. His nickname is a reference to rat droppings found in his pre-war Bugatti ("petoules" in French) which he used in the first post-war motor race in Europe.

MICHELE ALBORETO, 1981 to 1994

Italian drivers were generally typecast as wild, crazy, dramatic and overly emotional and Alboreto was anything but that. That is why he fit perfectly in the **Tyrrell** team in the early part of his career, and was hired by **Ferrari**, staying there from 1984 to 1988. At Tyrrell he had a learning curve year in 1981 but started to score points early in the 1982 season, getting his first podium at **San Marino**, a race that was boycotted by most British teams. He scored points in four more occasions, the high point of the season a win in the last race at **Las Vegas**. Michele continued at Tyrrell in 1983, and won at **Detroit**, the last normally aspirated engine win until 1989. Fighting for top results against the turbos was impossible, so he scored only one additional time. His big chance came at **Ferrari**, 1984, where he had his best years, in fact, it was the only truly competitive car he ever drove. He won a race in 1984, Belgium, and then a couple more in 1985, finishing the year runner-up to **Alain Prost.** Alboreto would remain at

Maranello an additional three seasons, but although he scored many times, with eventual podiums, he never won again and his best championship result after 1985 was 5th in 1988. He returned to Tyrrell for a short while, where he got his last podium, a 3rd place in the 1989 Mexican Grand Prix, finishing the season at **Larrousse.** Three seasons at **Footwork** were mostly disappointing, including the dreadful **Porsche** V12 engine used in the early part of 1991. As many Italian drivers of the age, Alboreto did stints at both **Scuderia Italia** (1993) and **Minardi** (1994), where he scored his last point from 6th in Monaco. He then retired from the category, becoming a sports car driver and even dabbling in Indycars. He won **Le Mans** with **Kristensen** and **Johannson** in 1997 and died testing an **Audi** sports car in 2001.

GRAHAM HILL, 1958 to 1975

Graham Hill enjoyed a very long career in Grand Prix racing, most notably driving for **Lotus**, where he started, and **BRM**, in which he won his first championship. The first and still only driver to win the World Championship, the **Indy 500** and the **24 Hours of Le Mans**, Graham also raced extensively out of F1. He is widely reckoned to be **Jim Clark's** main competition in the 1.5 liter years, and joined the great Scot at Lotus for the 1967 season. When Jim died in a F2 race in 1968, Graham provided the necessary leadership to Lotus, and won the 1968 title. A late season 1969 accident affected his competitiveness somewhat, but Graham continued racing in private Rob Walker's Lotus, **Brabham, Shadow, Lola** and his own **Hill**. After officially retiring in 1975, Graham looked forward to a long and successful career as a team owner and constructor but it was not to be – a plane crash claimed the lives of Graham, driver **Tony Brise** and other team members.

JOAKIN BONNIER, 1956 to 1971

The bearded Swede drove in Formula 1 for a long time, and is best known for winning **BRM**'s first GP in 1959. He also led a few other Championship races, and spent most of the last years of his F1 career driving older, less competitive machinery. A surprise entry in a Honda in the 1968 Mexican Grand Prix showed he could still be competitive, finishing 5th. He retired from F1 after the 1971 season, but continued racing his beloved sports cars. He died at **Le Mans**, 1972, when his Lola collided with a slower **Ferrari**. A sad end to one of the sport's safest drivers.

ROBERT KUBICA, 2006 to current

Modern Polish Grand Prix drivers are a rarity, in fact, Kubica is the only one to have made the cut. A sublime talent, Robert raced in myriad Formula **Renault** championships starting in the year 2000, finally winning the European series in 2005. That spurred the interest of **BMW Sauber**, who gave him rides for the last six races of the 2006 season. He retributed the favor qualifying in the top 10 in all but one race and finishing 3rd in **Monza**. He was obviously retained by the team for 2007 and finished a large number of races in the points, the season best placings a trio of 4th places. He also had a horrifying crash in Canada but went back racing right away. His one and only Grand Prix victory came in the same place, Montreal, 2008. He also scored 6 other podiums, finishing fourth in the Championship. The 2009 season was far from excellent, but there was still a podium, and for 2010 he moved over to the Renault camp. Things looked fine, he still got the occasional podium and finished 8th in the championship. He had an early season rallying accident at **Andora,** Italy, and was trapped in the car for over an hour, and the extent of his injuries was very substantial, including the partial amputation of his forearm. That was widely reckoned to spell the end of his Grand Prix career.

However, Robert first worked on his recovery, then went back rallying, raced wherever he could in many categories, never losing faith. In 2017 he was given a test by Renault at **Hungagoring**, and finished a remarkable 4th. Kubica was hired as a **Williams** test driver for 2018, and quite amazingly, was given a race seat by the team for the 2019. Although Williams was not a dream ride at the time, Kubica soldiered on, and years after his accident, got one point from tenth, in Germany. He was then hired by **Alfa Romeo** as a test driver and got two starts in the 2021, driving in Holland and Italy. He is still a test driver, so, who knows, may drive yet again. Quite an amazing story.

JACK BRABHAM, 1955 to 1970

A 3-time World Champion, Black Jack had a very unusual career, in that he raced for many years, but his wins came in leaps and bounds, in only 5 seasons, 1959, 1960, 1966, 1967 and 1970, winning many races in a row on two occasions (5 straight races in 1960, 4 straight races in 1966). He helped bring the rear-ended F1 car revolution, winning the 1959 title for Cooper, then began to field his own Brabham cars in 1962. He became the first F1 driver to win a race in his own car, and only one to win a championship under such condition. In the year of his retirement Brabham was still competitive, winning the first race of the year, and losing the Monaco and British Grand Prix on the last lap, both times to future champion Rindt.

NIKI LAUDA, 1971 to 1985

Niki Lauda`s two first season's driving Marches in F-1 were not indicative of the driver that blossomed driving for **Ferrari.** He already showed a stroke of genius in his 1973 season at **BRM,** but the real speed and skill became obvious right at the start of the 1974 season, when he finished 2nd in Argentina.

Lauda rose to prominence in 1974 (Alejandro de Brito)

The first wins came soon, and more reliability gave Lauda the 1975 championship. He was likely to win again in 1976, but a fiery accident at the **Nürburgring** changed the dynamic. He did not win the championship, but won everybody's hearts returning to racing a little over a month after being given his last rites. He stayed at Ferrari one more season, won another title, then moved over to **Brabham**, where he made a lot of money with scant results. He then retired for the first time, to run his airline Lauda Air. Success in the air was not as easy as on the racetrack, and Lauda came back to racing in 1982. He stayed on until 1985, and won a third title in 1984, by half a point over **Prost**. He continued his involvement in racing, in fact was part of the **Mercedes** team, in a managerial role.

JOHN SURTEES, 1960 to 1972

John Surtees was and remains the only driver to be a World Champion in both motorbikes and cars. He first came into the sport in 1960, driving for Team **Lotus**, finishing 2nd in his second race at Monaco. He shifted to Yeoman Credit/Bowmaker, running **Coopers** and **Lolas**, and got two second places in 1962. His move to **Ferrari**

came in 1963, and his first win was in the Nürburgring. Big John got his championship in 1964, winning two races, and some tricky teamwork by Ferrari assured he beat **Graham Hill** and **Jim Clark** for the title. He continued at Ferrari until the 1966 Belgian Grand Prix, which he won, then left the team over disagreements. At Cooper he won another race and finished second in the Championship. Two years were spent at **Honda**, where he developed further his car design skills, and a last win came at Monza, 1967. He raced for **BRM** in 1969, then started running **Team Surtees** cars in 1970, at first a McLaren, then the Surtees TS7. He got his last points in the 1971 season, but squeezed in a last start in Italy, 1972, which also happened to be the constructor's best result ever (**Hailwood**, 2nd). He continued as a constructor until 1978, but business and health reasons forced him to leave Grand Prix racing.

RUBENS BARRICHELLO, 1993 to 2011

When Barrichello first came into F1 **Ayrton Senna** was still around and winning. Brazilians had gotten spoiled; from no representation in the 60s, Brazilian drivers won no less than 8 championships, from 1972 to 1991. So a great portion of the country had gotten the misguided notion that it would produce champions forever, which has not been the case. When Ayrton died at **Imola**, 1994, the pressure on Rubens was tremendous right off the bat, but it soon became clear that Barrichello could be expected to win races, maybe a championship, but would not become a driver of Ayrton's stature. So he spent the 90s driving for **Jordan** and **Stewart**, midfield teams that would give him little to no chance at stardom. The big chance came for the 2000 season, at Ferrari, and Rubens did very well, even though he is widely criticized in the country for playing second fiddle to Schumi. There he won nine races, was runner-up in the Championship and scored dozens of podiums, plus a nice helping of poles and fastest laps. He was hired by Honda for the 2006 season, and his career seemed in trouble when Honda bowed out in the end of the 2008 season. The team was sold to **Ross Brawn**, who retained both drivers, and Rubens had the chance of driving for a fast team again. He won a couple of races in that team, and finished 3rd in the Championship. Brawn was sold to Mercedes, who had other thoughts about driver choice. Rubens went to Williams, replacing one of the drivers who was hired by Mercedes, **Nico Rosberg**. His seasons at Williams did not yield great results, except for a pole, and sensibly Rubens called it a day at the end of 2011, in F1, that is. He continues to race competitively in the Brazilian Stock Car Series. He can proudly boast 11 wins, 68 podiums, 17 fastest laps, 14 poles, and 323 starts.

JENSON BUTTON, 2000 to 2017

Despite some adversity, such as a so-so rookie season in 2000, and then a dreadful time at **Benetton**, 2001, **Jenson Button** managed to have a very long Grand Prix career, winning 15 Grand Prix and the 2009 World title. His career in lower formulae was short and reasonable, rather than stellar, and many felt he was out of place coming into F1 at 20 years of age. Perhaps another season of F3 was in order, but he refused to do it. His first F1 win would come only in his seventh season, by which time most Grand Prix careers are over, yet, he finished the 2004 championship in 3rd place driving for **BAR Honda**, a team that looked hopeless until then. His first win came at Hungary, 2006, driving a Honda, and his championship season seemed surreal. He drove for a rookie team, **Brawn**, and won most races of the first part of the championship. Then it was just a matter of finishing in the points, and walking away with the crown. The rest of his career would be spent at **McLaren**, where he got 8 further wins, finishing runner-up in the 2011 season, then endured the dreadful **McLaren-Honda** seasons of 2015 and 2016, doing one final race in 2017. He was known for a very polished and car sympathetic driving style.

KIMI RÄIKKÖNEN, 2001 to 2021

Kimi Räikkönen was quite the character in his F1 years, but he knew what he was doing (Alejandro de Brito)

The Finn had only 23 car races under his belt when he was hired by **Sauber** for the 2001 season. By 2003 he was a **McLaren** driver, where he got his first wins, and got two runner-up positions in the Championship. A move to **Ferrari** in 2007 yielded his sole championship under very competitive conditions, then he dropped out in the 2010 season, for Ferrari wanted **Alonso** and paid Kimi not to race. It would not be the first time a Finnish driver ended his F1 career early (both **Keke Rosberg** and **Mika Häkkinen** retired way too early). Kimi then spent a couple of seasons driving rally cars, NASCAR trucks, but had unfinished business in F1: he was convinced back by Lotus. Kimi had a wonderful season in 2012, finishing 3rd in the championship and winning at Abu Dhabi then at Australia, in 2013. Formula One is a small world, and then Ferrari wanted him back for 2014, to partner Alonso! His second foray at Ferrari yielded a single win, but a large number of podiums, tons of fastest laps (an area in which Kimi excelled) and Räikkönen, known for disconcerting one-liners, became the most popular F1 driver. He finished

3rd in his last Ferrari season then went to **Alfa Romeo** for three seasons, in essence, returning to **Sauber**. He did what he could with limited equipment, and by the end of 2021, retired from F1. The last driver active in F1 who was born in the 1970s, Kimi is set to try NASCAR again. Maybe Ferrari will call him a third time?

FERNANDO ALONSO, 2001 to current

The Spaniard was the youngest champion ever when he won his first title in 2005, so it is a bit funny to see him as the eldest statesman in Formula 1. Fernando is reckoned by some the most complete driver of his generation, even though his record does not reflect such stature. Humble beginnings at **Minardi** were followed by four seasons and two championships at **Renault.** A move to **McLaren** in 2007 was disastrous for the team, as Alonso and **Hamilton** fought for the team's attention but the two number 1 drivers concept never worked in Formula 1, in any era. So he came back for Renault for a couple more seasons, but the magic was gone. Ferrari wanted Alonso, trusting he could bring Maranello a few titles and perhaps end his career there, as **Michael Schumacher** said he would. Fernando spent five seasons at Ferrari, was runner-up to **Vettel** a couple of times, but the team did not get the hybrid concept right, and his 2014 season was just average. His last six seasons in F1 (four at McLaren, 2015 to 2018) and two at **Alpine** (2021 and 2022) have been disappointing. He can still drive exceptionally, but the wins, podiums and championship challenges are all gone, having led only two laps since leaving Ferrari. His early season at **Aston Martin** in 2023 has been outstanding, so who knows, Alonso may still win a couple of races. He did manage to win **Le Mans** a couple of times, driving for **Toyota** and came close to win the Indy 500, almost equaling **Graham Hill's** record.

LEWIS HAMILTON, 2007 to current

Hamilton was very young when he became a **McLaren** driver in 2007, sharing duties with **Alonso**. McLaren could not really decide which driver it would support, so the title was lost to **Kimi Räikkönen**, who had a great end of season. The first title came soon, 2008, but although he won on occasion during the Brawn season and **Vettel**'s era of domination, it became clear that McLaren would not be a good option on the long-term, so he left for the 2013 season, going to **Mercedes**. Rather amazingly, Lewis has stayed this long in F1 driving only cars with Mercedes engines, so he is the quintessential company man in F1. At Mercedes Lewis did wonderful, won six other titles, and had double digit wins in six of these seasons. He has basically beat most records (wins, poles, podiums, laps in the lead, km in the lead, points), and only fastest laps and most wins in a season (Michael Schumacher got 13 in 2004, and Vettel repeated the feast in 2013) have eluded him. He is still very fast, but the 2022 season has not gone his way, so far.

SEBASTIAN VETTEL, 2007 to 2022

When he debuted in 2007, driving a **BMW**, Vettel became the youngest driver to score a point at the time. Then he immediately moved to the **Red Bull** set up, at first at **Toro Rosso**, the junior team. The young German embarrassed the senior team by winning a race for Toro Rosso before **Red Bull**, at **Monza** 2008, so he was quickly moved over to Red Bull for the 2009 season. There he was the main challenger to **Brawn Grand Prix**, finally winning Red Bull's first race, and finishing runner-up to **Jenson Button**. He then went into a rampage, won four titles in a row, between 2010 to 2013, winning 34 races. In 2014 things did not go well, and a pattern emerged: Vettel seems to get intimidated by fast younger teammates, and **Ricciardo** won 3 races, to Seb's zero. He was scooped up by **Ferrari**, as a replacement for **Alonso**, and in his period there managed a few wins and a couple of runner-up spots to Hamilton. The arrival of **Charles Leclerc** in 2019 seemed to unsettle Vettel somewhat, so after two seasons running behind the Monegasque, Vettel went to **Aston Martin**, and retired at the end of the 2022 season as no progress was made by the team and the German. Then came 2023 and **Alonso**...

MICHAEL SCHUMACHER, 1991 to 2012

A **Mercedes** protégé in his early days, Michael was hired as a replacement for **Bertrand Gachot** at **Jordan** in the Belgian Grand Prix of 1991. His debut is somewhat overrated, for he qualified 7th and did not complete a lap. The Jordan was not a bad car, for De Cesaris, Gachot and then Moreno placed the car in similar areas of the grid during the season, so Schumi did not do anything special at Spa. But it was a now or never situation: the first to grab the German would probably have him for a few seasons, and Mercedes was not yet around. Ultimately **Benetton** hired him and fired **Moreno**, who drove the next race for Jordan and never again had a competitive F1 car. Tough break, but an understandable situation. Michael immediately proved fast, even though he would only score 4 points in that initial short season. He remained at Benetton for four seasons, and became a constant points scorer in 1992 and 1993, winning one racing a piece in these seasons. For 1994 he was already experienced, and Benetton finally got a good car: Michael beat now **Williams** driver **Ayrton Senna** fair and square in the first two races of 1994. After the Brazilian's death at **Imola,** Michael was obvious favorite for the title, although **Damon Hill** held up Williams' honor admirably. Michael won 1994 and then 1995, by then with the excellent **Renault** engine. At the end of the season he left for **Ferrari**, but the first four seasons at Maranello were not easy. He was runner-up in 1998, but his points were removed in 1997 over his aggressive tactics in a championship decider. An accident sidelined him for some races in 1999, so he closed the decade with a 5th place in the Championship. In 2000 Michael began a wonderful run of five straight championships driving for Ferrari, and beat every important record. After his final title, 2004, he stayed a couple more seasons at Ferrari, and retired. Mercedes managed to lure the German back after a 3-year sabbatical, to spearhead its return to Formula 1 competition. In hindsight, this was not a good idea. **Nico Rosberg**, his teammate was obviously faster than him just about anywhere, won the first race for Mercedes and one pole, one fastest lap and three laps in the lead was the best Michael could come up with. At the end of 2012 he retired for good, but a ski accident would take the great champion away from the public's eye.

RICCARDO PATRESE, 1977 to 1993

Having won the European F3 championship in 1976, Riccardo Patrese was hot property in the 1977 driver's market. He was hired by **Shadow** to replace another Italian, **Renzo Zorzi**, and had a typical learning-curve year. By the end of the season Patrese was qualifying in the top ten and ended the year with his first point. In 1978 most Shadow key personnel defected from the company joining the new **Arrows** outfit. In the second race, in South Africa, Patrese surprised everyone leading the race on speed for 36 laps. The team was then sued by Shadow, who claimed the Arrows design had been stolen from Shadow, and lost the lawsuit. In the meantime, Patrese continued to qualify in the top 10, and finished 2nd in Sweden. The Italian was then involved in more controversy, accused of having caused the accident that killed **Ronnie Peterson** in Italy, and was banned for one race. Patrese would stay at Arrows an additional three seasons, and even scored a pole and led at Long Beach, 1981, before leaving for **Brabham**, a established team.

The first win came at Monaco, 1982, in a very confusing race where many frontrunners crashed in the latter stages.

Another year at Brabham yielded a second win, but less points, and he left at the end 1983.

Patrese spent four seasons at Arrows, where he occasionally showed a lot of speed, but had no wins (Kurt Oblinger)

The next berth was in the Euro Racing **Alfa Romeo** outfit, but after scoring 8 points in 1984, Patrese went scoreless in 1985. He returned to Brabham in 1986, and his situation was only slightly better, scoring two points in 1986 and six in 1987. Apparently past his time, Patrese then had a good chance in the end of the 1987 season, replacing the injured **Mansell** at **Williams**. Sir Frank liked what he saw, and the Italian was retained for the 1988 season. Just when one thought Patrese's career was about to fizzle, he began his best period in F1. Overall he stayed at Williams until the end of 1992, finished the World Championship in 3rd in 1989 and 1991, and then was runner-up to teammate Mansell in 1992. He scored most of his points, wins, podiums and fastest laps during his Williams years as a clear number 2 driver, but in the end of the 1992 season Sir Frank let both of his drivers go, who had finished 1-2 in the Championship. Patrese would enjoy one final season in F1 in 1993, as number 2 to Michael Schumacher at **Benetton**, finishing 2nd in Hungary and 3rd in Britain, then bowed out of the sport. A long and atypical career indeed.

THE GRAND PRIX THAT DID NOT HAPPEN

A huge number of auto races fall into complete obscurity the moment the checkered flag is waved, despite the effort and expense of the participants, often involving great sacrifice. Try for example to find the complete results of most races on the internet, or even magazines. With luck you will find major events, that is all.

The omission of race reports is not uncommon. What is unusual is to find a story about a race that did not happen!

Only one pair could get away with such mischief, Motor Sport Magazine, and journalist **Denis Jenkinson**, both British. Motor Sport is the oldest motoring magazine in the world, published since 1924, and, indeed, still exists today. Jenkinson is one of the best known, and some would say, best and most talented journalist specializing in motorsports. Author of many books, Jenkinson was also known for being the co-driver for **Stirling Moss**, with Mercedes-Benz in the 1955 Mille Miglia, duly won by the duo.

One of the peculiarities of Jenkinson was that he was "old school". It is true that in his later years, Jenks, as he was known, had already surrendered to modernity. But in the mid-70s, Denis, who signed his reports DSJ, resisted some winds of change that swept through motorsport. There was a famous exchange of "niceties" between Jenkinson and **Jackie Stewart**, on the very pages of Motor Sport in 1972. Jackie, who had a column in an English newspaper, strove to increase the safety of tracks, cars and racing, while Jenkinson believed that there were overreactions in certain corners, which yielded a nasty response from Stewart, published in the journal.

It is therefore not surprising that Jenkinson did not swallow the change of venue of the German Grand Prix, from the **Nürburgring** (Norsdschleife) to **Hockenheim** from 1977 on, largely due to the terrible accident that befell **Niki Lauda** in the 1976 GP. For Jenkinson, Nürburgring was sacred.

This was reflected in the September 1977 issue of the magazine. In addition to reports on GPs from Germany and Austria, a report appeared on page 44 entitled "Der Grosser Preis von Deutschland", which means German GP in German. In the index, one notices something peculiar - a reference to "Hockenheim Formula 1 Race", instead of the German Grand Prix. The title in the story itself is "Der Kleine Preis von Deutschland" (Small Prix of Germany), which indicates DSJ's obvious displeasure with the new location of the German GP.

It happens that DSJ, who did not sign the article "Der Grosser Preis von Deutschland", but whose style is obvious, used three pages of the magazine to chronicle the events of a F1 race that was never run! He did so in such a realistic fashion that some might insist today that the race indeed took place.

In the fertile mind of Jenks, two F1 races were held in Germany that year, the official held in Hockenheim and the popular Nürburgring race that showed traditional motorsport's resilience, surviving unscathed through the changes of the era. Thus was created the GP did not exist.

The narrative is delicious. Some clear Jenks preferences emerge. Among others, **Chris Amon** is called by **Shadow** to race one of their cars, but refuses, confirming retirement. Niki Lauda remains home, mowing the lawn. **Max Mosley** produces an overwhelming number of **Marches** for the race, as not all regular teams attended. **Tyrrell**, for example, did not field cars, although the long-retired Jackie Stewart offered to sort out the difficult six-wheel cars. And the six-wheeled March almost runs with **Ian Scheckter,** whose brother Jody decided to stay home, because the race did not count points towards the World Championship. **Dieter Quester** almost took part in a March. **Emerson** and **Copersucar** were testing, as always.

Here's the imaginary grid GP, with formation 3-2-3 (not used since 1973) and 22 cars:

1. Mass (McLaren)
2. Stuck (Brabham)
3. Hunt (McLaren)
4. Ickx (March)
5. Nilsson (Lotus)
6. Laffite (Ligier)
7. Stommelen (March)
8. Andretti (Lotus)
9. Derek Bell (March)
10. Regazzoni (Ensign)
11. Jones (Shadow)
12. Reutemann (Ferrari)
13. Jarier (Penske)
14. Tambay (Ensign)
15. Tim Schenken (March)
16. Ertl (Hesketh)
17. Merzario (March)
18. Schuppan (Surtees)
19. Lunger (McLaren)
20. Neve (March)
21. Edwards (BRM)
22. Henton (BRM)

In the sublime imagination of Jenks, **Tim Schenken** comes back to F1. **Ickx, Stommelen** and **Bell**, despite the March, are Top 10 on the grid. For Jenks, March was obviously the last chance to keep the independents in F1, and in his drama, the cars have a wonderful performance. In fact,

Ickx, an expert on Nordschleife, starts fourth, and runs towards the front for much of the "event". Jenks also dreams about two **BRM** on the track, a nice idea that last happened in 1974, but an obvious impossibility in 1977. With a dose of realism, the BRMs occupy the last row ...

Stuck took the lead with **Brabham**, but eventually loses his place to the two McLarens. Both BRM leave in the first lap, and **Brian Henton** almost drives away on his private March to continue in the race ...

In the end, **Jochen Mass** wins, delighting the German crowd, and **Hunt** beats the lap record of the Nordschleife.

The final result of the GP that never was:
1. Jochen Mass, McLaren M26
2. J. Laffite, Ligier
3. C. Reutemann, Ferrari
4. C. Regazzoni, Ensign
5. D. Bell, March
6. P. Tambay, Ensign
7. M. Andretti, Lotus
8. B. Lunger, McLaren
9. J. Hunt, McLaren
10. A. Jones, Shadow
11. V. Schuppan, Surtees.

The final comic note is the disappearance of **Vittorio Brambilla**, who left the track in the Surtees during testing. Big John was so busy with the new pupil **Schuppan**, that he did not notice the disappearance of the Italian. He had fallen down a ravine, and was trying to get the car out of there by himself, for two days.

For obvious reasons there are no photos of the event in the "report". The only three photos are from the 30's, illustrating **Caracciola**, a Mercedes and Auto-Union in the pits. A brilliant piece of sarcasm, if you ask me.

MAJOR HEARTACHES

STEWART FAILING TO START HIS 100TH GRAND PRIX AND PROPERLY RETIRE

It is normal nowadays for F1 teammates to hate each other's guts, sometimes for very little apparent reason. I suppose it shocks recent followers of the sport that some teammates were actually friends and supported each other. Such is the case of **Jackie Stewart** and **Francois Cevert**, at **Tyrrell**. When Francois joined the team, in 1970, Stewart was already a World Champion and a driver of major stature. Their relationship grew with time, and Stewart did not mind at all teaching the Frenchman all tricks of the trade. Cevert was 3rd in his first full season, but knew his place: foremost, his role was to support Stewart, in races and in the championship. In 1973 Stewart drove very well, won five races, and here and there Cevert appeared the faster man. Francois was in no rush, he knew his day would come.

Cevert's only Grand Prix win took place in **Watkins Glen**, in 1971. That was the venue where Stewart, who had just won his third world championship at **Monza**, was supposed to race in his final Grand Prix, the 100th one, and properly celebrate his title and pass the torch to his friend.

Unfortunately, Cevert had a terrible accident in practice, his body cut in half, and died on the spot. Tyrrell for obvious reasons withdrew Stewart's and **Amon**'s entries, and what should have been a festive day and perhaps Cevert's first win of the season, ended up in mourning.

ANDRETTI AT MONZA, 1978

1978 was finally **Mario Andretti**'s year in Formula 1. He had been around, on and off, since 1968, but his second stint at Lotus proved to be magic. In the **Lotus** 78 and 79 Mario proved to be the driver to beat in the 1977 and 1978 seasons, winning 10 races: his celebration day would obviously come at **Monza**, in the land where Mario was born 38 years earlier. Additionally, he could share the spoils with **Ronnie Peterson**, who was possibly the fastest driver of his generation, but also universally well liked. Both **Emerson Fittipaldi** and Mario Andretti said Ronnie was the best teammate they ever had, in two separate interviews. Ronnie did not mind playing second fiddle to Mario in 1978, after all, the American developed the cars.

In the end, Monza 1978 turned out to be a forgettable day for Andretti. Ronnie had an accident at the start and was taken to the hospital. The race continued, and Andretti and Villeneuve finished 1-2, but were punished 60 seconds for jumping the start, so **Lauda-Watson** were promoted to the top positions. Certainly a bit ticked-off, the worst was to come yet. Peterson had serious leg injuries, and in the hospital, had complications and went into a coma. A few hours later Mario's friend was dead.

JACK BRABHAM'S 1970 FOLLIES

Jack Brabham was already 44 years old, had won three world championships and intended to retire in 1970. But Jack didn't do things in half measures, so he started the season well, winning the South African Grand Prix. He remained reasonably competitive all year long, getting a pole and fastest laps, but lost two races on the last lap, both times to **Jochen Rindt**. He was leading the Monaco Grand Prix, when Jochen Rindt pressured him into making a rare mistake. It worked, Rindt passed him and Jack finished second. Then, at **Brands Hatch** he led into the final stages of the race, but ran out of fuel on the last lap and finished 2nd. The additional points he would get from wins would not put him in contention for the title, but perhaps, the success could increase his motivation going to the final stages of the season. In the end Jack ended 5th

in the Championship, while his rival Rindt won the championship. Brabham managed to retire in one piece.

MASSA LOSING THE 2008 CHAMPIONSHIP ON THE LAST LAP

We have gotten very used to seeing **Lewis Hamilton** winning the F1 championship in commanding style, forgetting he won his first title in 2008 by one point, in the last corner. After the 2007 internal battles at **McLaren**, the team had decided that a clear number 1 designation was better, so Lewis was number one in the team, after **Alonso** left. He won the first race of the season, but a challenge was expected from Ferrari, which won the second GP with **Räikkönen**. Ultimately the challenge did not come from the 2007 champ, but from Brazilian **Felipe Massa**. By the Italian Grand Prix Massa had five wins to Lewis' four, but the Brit had one more point than the Brazilian. The next race was held at Singapore, and Massa began on the right track, starting from pole, while Lewis started second. Massa retained his advantage over Lewis, but then **Fernando Alonso** pitted early to change tires. A couple of laps later, Alonso's teammate **Piquet** crashed, so the Spaniard gained an advantage. A messy Ferrari pitstop dropped Massa out of contention and in the end, Massa finished 13th and Lewis third. Things still seemed to go Lewis way, as he won the Chinese Grand Prix, while Massa finished 7th in Japan and 2nd in China. The decision would be in Brazil.

In his home race Massa did everything he had to do. He had pole, he had the fastest lap, he won, and his teammate Kimi was helping where he could. Lewis started fourth, but rain would still play a part in the event. All leading drivers changed to intermediate tires as it began to rain in the late stages of the race, except for **Timo Glock**. **Vettel** and Hamilton were battling it out, and passing Glock in the final corner was not that difficult, in fact, it looked easy given the German's wrong tires for the situation. As a result Hamilton got enough points from 5th to win the title by one point. Brazilian fans obviously thought that Glock made it easy for Lewis, and a few conspiracy theories were cooked. Had Lewis finished 6th, he would have tied with Massa, who had more wins. A bittersweet day for the Brazilian, who would never again have that chance. In fact, that was his last F1 win, although he would continue in F1 for quite a few more years.

FAILING TO START FROM POLE

Jean-Pierre Jarier was still in the beginning of his Formula 1 career in 1975, having driven two full seasons and done one race in 1971. He had done great in Formula 2 in 1973, in a March, and in Sports Cars in 1974, driving for **Matra**, but except for a podium result at Monaco, 1974 (3rd place) he had not shown much form in F1. Then came Argentina, 1975. The **Shadow DN5** looked and ran great, and Jarier was quite confident. The Frenchman scored an unexpected pole in a very competitive field, but he stripped his clutch in the warmup and would not start, a rare occurrence in the history of F1. Proving the **Buenos Aires** speed was no fluke, Jarier then scored another pole at **Interlagos**, the second race, and after running behind Reutemann in the early stages, passed the Argentine. He was running away with the race, when a fuel metering unit failed and he retired with a few laps to go, gifting the win to **Pace**. The season looked promising, however, the only point scoring performance Jarier had that year was 4th in the half race at **Montjuic**, so the pace was not sustained.

CHRIS AMON AT CLERMONT-FERRAND, 1972

Widely acknowledged the unluckiest driver in F1, the New Zealander was obviously at a breaking point in 1972. He had left **Ferrari** just before the 312B had become the car to beat in 1970, went to **March** believing a Cosworth engine was required to win, and then left the team for the 1971 season, the only season a March driver was a runner-up in the World Championship. Joining Matra, which ran its own 12-cylinder engine seemed a 180 degree change of opinion, but Chris followed his instincts, which granted, had failed him in the recent past. He managed to win his first race for Matra, the 2-heat, non-championship Argentine Grand Prix of 1971, and then started the South African and Spanish races on the front row. Things went downhill from there, until he got pole in **Monza.** He led there, as he had led many races before, but a change of tear-off visor gone wrong made him drift away from the leading group. In the 1972 season things did not start on the right foot but in Spain he began qualifying in the front pack, pitting for fuel when lying 3rd in Belgium. Then came France. A revised version of the car, the MS120D, was provided to Amon, with engine and other improvements, but most important of all, a decision had to be made about Matra's future racing activities, for the constructor was affiliated with **Simca,** in turn, owned by **Chrysler Europe**, was not sitting on a lot of cash and had no outside sponsor. The French outfit had just won the **24 Hours of Le Mans**, and in essence, it had to choose between F1 and Sports Cars. Amon responded extremely well to the souped-up car. He scored pole, drove away from everybody, and it seemed, was ripe for his maiden GP win. Then he got a puncture, for Clermont Ferrand was a mountainous circuit with pebbles galore on the sides of the narrow track. Amon dropped to eighth after the stop, but did not lose hope. He began to pass car after car, and in the end of the race was near champion to be **Fittipaldi**. The poor guy finished third. It is said that during a meeting with Matra brass in the aftermath of the race Amon agreed that it made more sense for Matra to pursue sports cars! In 1972 he would still get a front row place at Monza, but from then on Amon's topline career was over. A chance at **Tyrrell** backfired due to **Cevert**'s death, and he allegedly refused an offer to drive the second Brabham BT44 in 1974, over concern for the people that supported his Amon Formula 1 project. A nice, considerate human being, in addition to being a heck of a driver. The cover of this book is a picture of Amon from that very same race.

WILSON FITTIPALDI AT MONACO, 1973

Having a World Champion brother can be tough if you are a racer yourself. Much more so if you race at the same time, and even more if you are older than your sibling. Wilson had all that baggage to carry along, but he ran his own career well, to the extent that no sibling rivalry apparently existed between the Fittipaldis. His best chance happened in 1973. He had raced for **Brabham** for most of 1972, and had a number of cars available to him, a BT33, a BT34 and then a BT37, and got a couple of 7th places. For 1973 he was number 2 to **Carlos Reutemann**, and scored his first point in Argentina. In Monaco he qualified an excellent 9th (Reutemann started 19th) and was mixing it with the big boys all day. He was running 3rd when fuel system issues dropped him down the order, the possible podium vanishing before his eyes. He would still get a couple of points from 5th at the **Nürburgring,** and would do another full season of F1 in 1975, racing his **Copersucar-Fittipaldi**. He moved over to managerial duties in 1976, but after the team folded in 1982, Wilson continued racing in local events.

A BITTERSWEET END FOR GERMANS

It may sound weird saying one felt sorry for the lack of German wins in Formula 1 one day, after all, since 1991 Germans have won twelve F1 titles, plus a very large number of races. From 1934 to 1939, **Mercedes** and **Auto Union** won many GPs, most by German drivers such as **Stuck, Caracciola, Rosemeyer, Von Brauchitsch, Lang**. After the war, Germans were not frequently found in the top positions of the new World Championship, even though **Karl Kling** did lead a few laps in a Mercedes in 1954 to 1955. A light at the end of the tunnel was **Wolfgang Von Trips'** 1961 season, when he won two races and came very close to winning the championship, instead, meeting his death at Monza.

Yet, a few good German drivers did find their way into Formula 1, and there were two of them in the 1975 Spanish Grand Prix, held at beautiful **Montjuic Park** in Barcelona. Proceedings were tense from the onset, as drivers protested the badly installed armco barriers placed around the park, threatening not to race. In the end, everyone took the start, even though current World Champion **Emerson Fittipaldi**, and **Arturo Merzario**, drove the first lap in very slow speeds and retired in protest. Everyone else took the thing seriously.

The two Ferraris eliminated each other from the front row, and a number of leaders, some of them unusual at the time, took turns in the lead. One of them was **Mario Andretti**, a former GP winner driving the recently introduced **Parnelli**. On lap 17 another surprise: a German, **Rolf Stommelen**, was in the lead driving a debuting **Hill-Ford!** Rolf was very convincing on the lead, and it was felt that perhaps he could cause a major upset, for he was dicing and prevailing over Pace on a Brabham. On the 25th lap, the German got off the track, hurt himself on the loose armco, and a few people were killed in the process. Four more laps were run, and another German, **Mass**, had taken the lead. **Ickx** would pass him on the 28th lap, only to be repassed by the German. The race directors decided to call it a day, and a German had finally won a Grand Prix again since 1961, on a day two Germans led and one got hurt. Everybody was given half points, and neither Rolf nor Jochen would get another chance to shine. Rolf eventually died racing a **Porsche** in the US in 1983, whereas Mass, after a long career, managed to retire alive

SHORTIES

Carlos Reutemann posted the pole for his first Championship race at **Buenos Aires**, 1972, and seemed on path to win his first Grand Prix at home in 1974, when the airbox started to come apart and then he ran out of fuel. He was the fastest driver of the early part of the season, so a win would come sooner or later, and it did, in South Africa.

Reutemann come close to winning his maiden GP in front of an adoring home crowd in 1974. It was not to be (Alejandro de Brito)

Likewise, **Ronnie Peterson** had managed to finish a winless runner-up to **Jackie Stewart** in the 1971 season, led driving the recalcitrant **March** 721 in Canada, 1972, and then moved to the competitive Lotus team for the 1973 season. He was fast right away, and by **Anderstorp**, Sweden, he had already accumulated 4 poles. He drove away from the field at his home race, but an unlucky puncture gave the race away to Denis Hulme in the last lap. Things would get better, as he won the next race, in France, from 5th on the grid. **Jim Clark** loved winning GPs and could care less for 2nd places. In fact, he had a single second place in his 72-race GP career. In Italy, 1967, Clark had pole and led in the early part of the race. He had to pit to change a tire, then began a memorable drive from the back, retaking the lead on lap 61. Unfortunately, his car ran out of fuel on the last lap, but he still got third! Maybe he could live with a second place that day.

UNUSUAL ENTRIES

I have already mentioned that in the early days of the Championship organizers had enough discretion to accept or reject entries at their whim, sometimes based on very untechnical reasoning and the need to beef up thin grids. That resulted in some very unusual entries in the history of the championship.

In the very first season, 1950, **Harry Schell** entered the Monaco Grand Prix in a car that was far from a F1 or even a F2 car, a **JAP** (motorcycle) engined **Cooper** with a 1.1 liter engine. He qualified next to last and was involved in the pile-up that decimated 9 cars from the race. He would spend the rest of his career driving more orthodox machinery.

Clemente Biondetti was known as a successful Mille Miglia driver, having won the races many times. In F1, Biondetti is known for having driven the only true **Jaguar** engine ever used in F1 (remember the early 2000s Jaguar F1 car engines were Cosworths). The 3.4 liter engine was mated to a Ferrari 166S sports car, and retired from its single outing in the 1950 Italian Grand Prix.

Peter Broeker entered his **Stebro** car with a Ford engine in the 1963 US Grand Prix, and finished 7th having started 21st and last. He finished many laps behind the winner, but in front of Bonnier. This is Canada's contribution as a F1 constructor, even though some sources insist on identifying Wolf as a Canadian constructor.

Rodger Ward had won the 1959 Indy 500, so one can assume that if he wanted, he might have arranged a proper entry for the first F1 race in the USA, held in Sebring, Florida, that year. Instead, he was entered in a

1.75 liter **Kurtis-Kraft** midget, with **Offenhauser** engine, which was an obviously uncompetitive car for that event. He retired.

A midget against giants, Ward's mount in 1959. The midget lost. (Unattributed photo)

In his first GP entry, in the 1953 German Grand Prix, **Edgar Barth** (Jurgen's dad) was still an East German, but by 1957 had defected to the West. He became a **Porsche** factory driver, and was entered in a Porsche 550RS which was, after all, a sports car. Nonetheless, the car was entered in Formula 2. He was again entered in a Porsche Sports car in the 1958 race, entered as a Formula 2, and finished 6th. **Carel Godin de Beaufort** also raced Porsche sports cars in Formula 1 events in the beginning of his career.

The early South African Grand Prix welcomed many locally prepared and built cars, which did not race anywhere else. Among these cars were a few **Alfa Romeo** engined cars, such as the Alfa Specials raced by **Peter De Klerk**, and the Cooper and LDS Alfa. The **LDS**, by the way, was locally produced. It was basically a copy of Cooper models, built by **Louis Douglas Serrurier**. LDS cars continued to run until the 1968 event.

As mentioned elsewhere, **Cooper** chassis were mated with a number of engines from 1950 to 1969. **BRP** fielded a **Borgward** engined F2 car in the 1959 British Grand Prix for drivers **Bristow** and **Bueb**, which won the F2 class but finished 10th overall. The car did not appear in other Championship races.

Then a question lingers, why didn't anybody prepare supercharged cars for the 2.5 Liter Formula? After all, 750 cc engines could be fitted with superchargers and used in the championship. Well, at least two such cars were produced, but did not race in the Championship, and were insufficiently competitive. **Berardo Taraschi** drove a supercharged **Giaur** in the 1954 Grand Prix of Rome (non-championship). He qualified towards the back and retired after five laps. The French also tried to build a supercharged 750 cc for Grand Prix racing. The cars were produced by **Rene Bonnet**, of Le Mans fame, and equipped with supercharged **Panhard** engines. The **DB** cars were driven by **Paul Armagnac** and **Claude Storez** in the Pau Grand Prix of 1955 and were in a slow class of their own with the Volpini of Alborghetti, posting times 12 seconds off the worst **Gordini**. Needless to say, neither the Giaur nor the DB ventured into the Championship.

EARLY BLOOMERS
DRIVERS WHO WON IN THEIR DEBUT SEASON

Do not read into this any type of qualitative assessment: after all, quite a few drivers on the list did not progress much in terms of career wins, and many of the top drivers did not win on their debut season. I know some might consider the list bogus, because some drivers just drove a few races in their debut season, such as **Michael Schumacher.** Criticism accepted, however **Emerson Fittipaldi** won on his fourth race, and in total, raced 5 times in the 1970 season. I am not considering **Fangio, Farina** and **Parsons**, all of whom won races in the 1950 season, because there was no preceding season. Fangio and Farina had raced in Grand Prix prior to 1950, so the only one I would consider here would by Johnnie Parsons, whose 1950 Indy 500 was his first.

So here are the early bloomers:

LEWIS HAMILTON - Lewis was competitive right off the bat in **McLaren**, in 2007. The team had won 10 races in 2005, then none in 2006. Lead driver **Kimi Räikkönen** moved over to **Ferrari** and won the Australian Grand Prix of 2007. Hamilton and **Alonso** were brought on board the venerable team at the same time, and their relationship went downhill as the season progressed. Lewis led his first race from laps 19 to 22, but finished 3rd in his debut. He would only win his 6th start in F1.

JACQUES VILLENEUVE - In comparison, Jacques did much better than Lewis. He started on pole in his first race, Australia, 1996, and actually led from the start, until being passed by teammate **Damon Hill** on the 30th lap, regaining the lead, but being passed again by Damon. He also won earlier than Lewis, the 4th start. Unfortunately for the Canadian, his career was all about the first two years, fizzling very fast. He did finish his debut season runner-up to Damon Hill.

JACKIE STEWART - Jackie debuted in the last year of the 1.5-liter formula as a teammate to **Graham Hill** at **BRM**. He was on the podium in his second race, got quite a few points during the season, and won in his 8th start, in the slipstream battles at **Monza**. He fought against three former champs that day, **Clark,** Graham Hill and **Surtees**, and passed his teammate on lap 75. He finished the season in 3rd place.

(Photo Rob Neuzel)

CLAY REGAZZONI - Two early bloomers in one season! Clay had a bit of a wild reputation until the 1970 season, when he became a more polished Formula 2 driver and won the championship. Ferrari was looking for a number 2 driver, and gave Regga a chance in Holland, 1970. He scored in his first two GPs so Ferrari kept him. He won at **Monza**, on a day where many drivers led, and finished a very impressive third in the Championship.

JUAN PABLO MONTOYA - A lot was expected of Montoya in 2001, who had won championships on both sides of the Atlantic, in Formula 3000 and CART. The Colombian was certainly very fast, but it took a good part of the season for him to settle in the **Williams** surroundings. Like Regazzoni his first win also took place in Italy, where he also had pole. After a few more seasons at Williams and a couple at **McLaren,** Montoya left

Formula 1 in 2006, unable to beat either **Schumacher** and **Alonso**. The best he did in the World Championship was 3rd a couple of times.

EMERSON FITTIPALDI - Emerson debuted halfway through the 1970 season, as **Lotus** was seeking a replacement for slow John Miles. He got points in his second race, then was a surprise winner on the 4th start, at **Watkins Glen**, assuming the role of team leader left by the deceased **Jochen Rindt.**

Emerson Fittipaldi won in his fourth Formula One start (Alejandro de Brito)

BRUCE MCLAREN - The young New Zealander began his career in 1959, as a number 2 in the **Cooper** team, and by the end of the season had proved his worth, by becoming the youngest Grand Prix winner at the time winning the US Grand Prix from 10th. He followed that up with a win in the very next Championship race, at Argentina, 1960, essentially getting half of his wins in that short time span of two races. He was runner-up in the 1960 championship, and by 1966 had become a Formula 1 constructor. He died testing a Can Am car in **Goodwood,** 1970, still very young, at 32.

GIANCARLO BAGHETTI - Not only did Baghetti win a Grand Prix in his first season, he also won his very first Championship race, the French Grand Prix of 1961. Ready for more? He had also won his first, non-championship race, the Syracuse Grand Prix. A new **Ascari** or **Nuvolari** in the making? Not so fast. After such scintillating debut Baghetti never came close to reproducing this form in F-1, and by the time he retired from F1 in 1967, he managed only five more points in the 1962 season. **Enzo Ferrari** had promised FISA, the Italian Federation, that he would field an Italian driver during the season, under the auspices of FISA. And so he did at Reims. The Italian overtook **Dan Gurney** in the last lap, and got his one Grand Prix and such an amazing record.

...LATE BLOOMERS
DRIVERS THAT TOOK A LONG TIME TO GET THEIR FIRST WIN

Most Grand Prix winners get their first GP win between the 2nd and 4th season, which basically seals their fate as a long-term F1 driver. Some have managed to stay in the game winning at latter stages in their career, some of them a long time.

One should not forget that the **Indy 500** was part of the World Championship from 1950 to 1960, and the one driver that took the longest to win was a driver that debuted in the 1950 Indy 500 and won the 1960 edition, **Jim Rathmann**. Jim, whose real name was Royal, took part in all Indys of the period, and quit after getting his win. A few other Indy winners also took some time to achieve their goals: **Jimmy Bryan** debuted in 1952, but only won in 1958, **Pat Flaherty** debuted in 1950, but won in 1956, **Sam Hanks** debuted in 1950 and won in 1957, and **Rodger Ward** almost matched Rathmann's record, debuting in 1951 and winning in 1959.

As of late, **Carlos Sainz Jr.** finally won his first race in Britain, 2022, having driven for **Toro Rosso, Renault, McLaren** and **Ferrari**. Carlos first came into the sport in 2015, as a teammate to **Max Verstappen**, and although highly rated, could never get around to this GP winning business until now.

Another late bloomer who finally won his first GP is **Sergio Perez**. The Mexican has been around since 2011, and has driven for **Sauber, McLaren, Force India, Racing Point** and now, **Red Bull**. His third season, at McLaren, was particularly nasty, for his best result was 5th. He stayed at Force India for many seasons, and continued after the team became Racing Point in 2019. While driving for this team Perez got his first win in the 2020 Season, driving from the back and finishing the championship in 4th place. This prompted his signing by Red Bull, where he has won additional races.

Some drivers are simply so good that teams hang on to them despite the lack of wins. Such was the case of **Mika Häkkinen,** who first drove for **Lotus** in 1991. He did very well considering the shambles the team had become, and in late 1993 was called by **McLaren** to replace **Michael Andretti**, getting a podium right away and giving **Ayrton**

Senna a fright. Then, after many seasons at McLaren, Häkkinen finally won his first Grand Prix, in 1997. That was followed by two straight titles, in 1998 and 1999, so it was worth keeping the Finn around.

A couple of other champions were also late bloomers. **Nigel Mansell** debuted in 1980, staying at **Lotus** until shifting to **Williams** in 1985, and would only start his winning ways in the late 1985 season. His championship would only come seven years later!

Nigel Mansell took a while to win his first Grand Prix, but in the end, bagged 31 of them (Alejandro de Brito)

Jochen Rindt debuted in F1 in 1964. He was very fast, would win in anything and was considered the undisputed king of Formula 2, but his Grand Prix rides, such as **Cooper-Maseratis** and **Brabham-Repcos** were insufficiently fast for wins. He did lead a GP as early as Belgium, 1966, and finished that season in 3rd place. He was hired by **Lotus**, in 1969, where he won all six of his Grand Prix, becoming posthumous World Champion in 1970. His first win was the late season U.S. Grand Prix of 1969.

Jenson Button is a case apart. Jenson did not have what we can call a stellar junior formulae career. At 19 years of age, he finished 3rd in the British F-3 Championship, and lacked the experience to land a F1 ride, but that is exactly what he did: a place in the **Williams** Formula 1 team, which just began a partnership with **BMW**. He scored points a few times, the best placing a fourth, and looked promising enough to be hired by another former champion team, **Benetton,** for the 2001 season. That turned out to be a disaster, as he got only two points in the entire season. He stayed on as the team became Renault, and then moved over to **BAR Honda**, getting 10 podiums, and finishing 3rd in the 2004 season on the strength of many point scores. His maiden win would occur only in his seventh season, in Hungary 2006, by which time BAR had become **Honda**. He would win 14 other races and stay around until 2017.

Rubens Barrichello was never a champion but was talented enough to ensure his long-term permanence in F1. He became a **Jordan** driver in 1993, and although he had scored a pole, fastest laps and even led a few laps, the team was pretty much second tier. A 3-year stint at **Stewart** yielded no wins, but he led a total of 67 laps in the 1999 season, which was enough to impress Ferrari. The Brazilian stayed in the Italian team from 2000 to 2005, and would only win his first GP in Germany, 2000, seven years after debut. He would still win many other races for **Ferrari** and a couple more driving for the **Brawn** team in 2009.

Giancarlo Fisichella also managed to stick around for a long time before getting his first win. The Italian first drove for **Minardi** in 1996, then moved on to **Jordan**, staying one season, then moved to **Benetton**, where he raced for four seasons. He then returned to Jordan, which had seen better days, but lucked out winning under freak conditions in Brazil, 2003. He would still get a couple of wins driving for **Renault** in the 2005 and 2006 seasons, and also drove for **Sauber** and **Force India**. He ended his GP career the worst possible way, disappointing greatly when **Ferrari** gave him an opportunity to fill-in for injured **Felipe Massa** in 2009.

Jarno Trulli is another Italian that took a long time to win. He actually impressed greatly in his debut season, 1997, when he led 37 laps in the Austrian Grand Prix, driving a **Prost**. Eventually he made his way into the **Renault** team, and won **Monaco**, 2004, leading 72 of the 77 laps. A remarkable qualifier, Trulli led a total of 16 races, but could not keep strong pace for entire race distances. He would often lead a trail of faster cars behind him, in latter stages of races, which became known as a Trulli train.

Jo Siffert only won his first Grand Prix in his seventh season in Formula One, but the Swiss has a better reason than most to justify this. Siffert spent all these seven seasons, starting in 1962, driving for privateer teams, such as Filipinetti, Ecurie Nationale Suisse and Rob Walker, in addition to entering cars himself in the 1963 and 1964 seasons, and when he did win his first race, the British Grand Prix of 1968, he was still driving for Rob Walker. In fact, Siffert only drove for works teams in his last two seasons in F1, **March** in 1970 and **BRM** in 1971. That is quite an amazing feat. (Photo Rob Neuzel)

Thierry Boutsen began in F1 in 1983 but only won in 1989, when he first came into the **Williams** team. In total he won three races, but was replaced by **Nigel Mansell** for the 1991 season and never looked competitive thereafter. **Ritchie Ginther** first drove in F1 in 1960, and drove for top teams **BRM** and **Ferrari.** However, he would only taste success in the last race of the 1.5-liter formula, when he won the 1965 Mexican Grand Prix in a **Honda.** Prior to this win Ginther had finished 2nd in no less than 8 GPs!

Peter Revson had briefly driven in Formula 1 in 1964, in the **Parnell** team, and would only win his first Grand Prix in 1973. That is a long time, however, Revson was out of F1 by the end of the 1964 season, and would only come back in 1971, driving a third **Tyrrell** in the US Grand Prix. So, in essence, he won in his fourth season.

BACK AND FORTH, FORTH AND BACK

Formula 1 has had a love-hate relationship with carmakers over the years, and in this day and age of excessive emphasis on marketing and corporate governance, some carmakers come and go from the sport with a nagging frequency, based on Board of Directors' mood and sympathy.

Alfa Romeo

The Italian automaker was founded in 1910, and by the mid-1920s it was fast becoming the top Grand Prix constructor of the world. It weathered the German constructors' juggernaut of the 1934-1939 period, winning occasionally, and survived World War II as a company. Alfa became one of the first teams to tackle racing seriously in the after-War period, and in 1950 it had a splendid car, the Alfa Romeo 158, which was a pre-war model that had been equipped with a supercharged engine. The car utterly dominated the Championship's initial season, and although **Ferrari** had become a threat in 1951, the Milanese company won both initial championships. At the end of the year it quit racing, for Ferrari had a newer model, and the latest 159 model would certainly become dated soon. Better to quit while on top.

The only Alfa presence in Grand Prix racing in the early 60s were Alfa Giulietta engined specials that raced in South Africa during the 1961-65 1.5 Liter era, that did not race elsewhere, and a **Conrero** tuned Alfa engine used in a **De Tomaso** during the 1961 season. They were not particularly fast, though.

In the mid-60s Alfa began to build fast sports cars and prototypes, and by 1969 it had a 3-liter, 8-cylinder engine that would be theoretically suitable for Formula 1. The company reached an agreement with **McLaren**, and a McLaren-Alfa was fielded in most Grand Prix of 1970 for **Andrea de Adamich**, and an additional car for **Nanni Galli**, in Italy. The car was basically much slower than the Ford version, so for 1971 the deal was shifted over to **March**. Both Galli and De Adamich continued to suffer with the lack of speed, even though in the prototypes the engine was particularly effective that season. By the end of 1971 Alfa was gone again.

Alfa Romeo came back and forth a few times. (Kurt Oblinger)

In 1973 Alfa began to develop a 12-cylinder powerplant for its prototypes, which proved faster and more competitive. The engines were tried by the **Hill** team during the 1975 season, but then an agreement was reached with Brabham, at the time one of the top teams. The partnership was mildly successful. The engines proved heavy and thirsty in 1976, improved in 1977, leading races, and **Niki Lauda** was hired by the team for 1978, his substantial retainer paid by Parmalat. The only two **Brabham-Alfa** wins took place in 1978, both under unusual circumstances. In Sweden Brabham had installed a fan on the back of the car, that was eventually ruled illegal, yet **Bernie's** team got to keep the win. In Monza, the top two on the road, **Andretti** and **Villeneuve**, were given 60 second penalties for jumping the start, and the win was handed to **Lauda** and **Watson**. Alfa's partnership continued in 1979, but in the end of the season the team had shifted back to Cosworth engines, because Alfa had decided prior to that to come back as a full-fledged works team, and cars were fielded for **Giacomelli** and **Brambilla**, starting in the Belgian Grand Prix in 1979.

In this second works team iteration Alfa Romeo stayed around from 1979 to 1985. It led races, got occasional poles, it even had former champion **Mario Andretti** in the driver roster in 1981, but never got around to winning a single race. In addition to the 12-cylinder 3-liter engine, Alfa also developed and raced a 1.5 turbo engine. In the last two years Alfas were run by **Euroracing** rather than **Autodelta**, and engines were also provided to the Italian **Osella** team, from 1984 to 1988. In this last season Osella called the engine Osella. **Ligier** was also supposed to use the Alfa engine, but the agreement was undone because driver **Arnoux** criticized the Italian engine.

From that point on Alfa concentrated on touring car racing, and by the early 2010s there were some fears that **FIAT** would discontinue the brand. Happily that did not happen, in fact, a strong line-up was introduced, and Alfa was again considered a premium brand in many important markets, including America. Mid-decade Alfa Romeo decals start appearing on Formula 1 Ferraris, opening the way for a return of the traditional brand to F1. That occurred in 2018, when the **Sauber** team was rebranded Alfa Romeo, with no change of ownership, the cars equipped with Ferrari engines! For the time being it seems to be a sustainable situation in the mid-term, but **Andretti** tried to buy the team, and Audi bought a minority stake in Sauber to enter F1 and may yet make an offer that cannot be refused. Don't be surprised if Alfa disappears again.

Renault

Renault was a major Grand Prix force even before Alfa's founding, in the first decade of the 1900s, but settled into a role as producer of sensible transportation for many decades. In the mid-60s Renault began to build a serious motorsport reputation, involving single-make series for its tourers, Formula 3, Formula 2, 2-Liter Prototype and Le Mans through its subsidiary **Alpine**. In 1976 it announced it would build a turbo Formula 1 car.

There have been a few iterations of Renault in F1. This is the original
(Kurt Oblinger)

The Renault F1 debuted in the British Grand Prix of 1977, and teething problems brought most retirements in the first year of operation. However, the car qualified 3rd in two Grand Prix of 1978 in Austria and **Monza**, so progress was achieved. The first pole would come soon in the heights of **Kyalami,** 1979, which favored the turbo engine, and the first win took place at **Dijon**, in the 1979 French Grand Prix. The team soon became a regular winner, and in 1981 French sensation **Alain Prost** was brought into the team.

Prost won many races and lost the 1983 championship by a scant margin, but then moved over to **McLaren**. In 1983 Renault had also began providing engines to Lotus, and then Ligier in 1984. Prost's replacements did not match *Le Professeur's* racecraft, and by 1985 the Renault works team was no more, and engines stopped being supplied to other teams after 1986.

Renault would not stay off the field for long. In 1989 3.5-liter Renault engines were provided to **Williams**, which won on the first year, and became stronger with every season. By 1991 the engine was as fast as the **Honda**, and by 1992, much superior. Thus, Renault won its first title in 1992, albeit as an engine supplier for Williams, a situation that was repeated in 1993, 1995 (as supplier to **Benetton**), 1996 and 1997. Then the bombshell: the Renault works would stop being an engine supplier, and the engines were rebadged and prepared by **Playlife** and **Mecachrome**. The magic was gone.

The Renault name would return once more, equipping the Benetton cars in 2001, and in 2002 the team was sold to the French carmaker, which became a works team again. The first win with **Alonso** came in 2003, the two and only titles as a works team in 2005 and 2006. Since then Renault has stayed as an engine supplier in F1, even though the same cannot be said of the Renault team. Its last season during this period was 2011, when cars were entered by the **Lotus Renault GP Team**, as the team fought for use of the Lotus brand in F1.

In the meantime Renault had become an engine supplier to **Red Bull,** a relationship that went really well from 2009 to 2013. With the change of engine regulations to the 1.6 hybrid turbo era, the relationship between the partners went into turmoil, as Renault was unable to produce engines that could match the **Mercedes.** The engines were rebranded **TagHeuer** in 2016, but continued to be called Renault at Lotus until 2015. In 2016 Renault discontinued the Lotus brand and restored the Renault nameplate in F1. That lasted until 2020, after which the team's name was changed once more, to Alpine!

Honda

Honda has also been in and out the scene many times. When it first started in Formula 1, in 1964, it was just starting to build cars, and was mostly known for its motorcycles, on and off the racetracks. The white Hondas soon got on the pace, and **Ginther** won the make's first race in 1965. **John Surtees** took the team to a different level, and won the 1967 Italian Grand Prix and finish the championship equal fourth with **Amon**. The Brit would stay on board one more season, but at the end of 1968 the Japanese carmaker retired from the sport.

Honda restored its love affair with global racing in the 80s, building Formula 2 engines that became the class of the field. As turbo engines became the wave of the future, Honda's interest increased, and by 1983 it was back in F1, as a supplier to **Spirit** which raced its engines in Formula 2 and then to **Williams**. The partnership developed well, and by late 1985 Honda was clearly the best F1 engine. Lotus would be added to the roster in 1987, and that same season Honda won its first championship as a powerplant supplier. A move from Williams to **McLaren** for 1988 meant four other straight titles for Honda, both turbo and normally aspirated engines. However, Honda's interest soon vanished and the nameplate disappeared, even though engines were still provided as **Mugens** to a number of teams from 1992 until the year 2000. The Mugen engine would win four races in Formula 1, three powering **Jordans**, one in a **Ligier**.

The Honda name would come back in the year 2000, in the recently formed **BAR** team. Despite the obvious enthusiasm, BAR would not win a single race as a constructor. The engines were also provided to Jordan, in

2001 and 2002, and in 2006 BAR became Honda. Engines were also supplied to a new Japanese team, **Super-Aguri**. The change of ownership did some good, as Button won his first GP at Hungary, 2006. The 2007 and 2008 seasons were not particularly good for Honda, and in the wake of the 2008 financial crisis, the Board decided to exercise austerity, announcing the end of the team. An agreement was reached with team manager, who took over the team and renamed it **Brawn**.

After a few years off, Honda came back once more as an engine supplier to McLaren in 2015. The dreams of restoring 1988 glory soon became a nightmare, as the engine was simply not very good and the performance sadly lacking. They stuck together for three seasons, and by 2018 Hondas had found their way on the back of **Toro Rossos**. There was visible improvement, to the extent that in 2019 the engines were equipping the senior team Red Bull. After a couple of learning seasons and a few sporadic wins (Red Bull initially won with it in Austria, 2019, while Toro Rosso also won in 2020, Italy) the Honda came good in 2021, after Honda announced it would pull out of the sport once more! Red Bull branded the engine operation, and Red Bull became an engine builder. Then Honda decided to come back… Can Honda ever make up its mind?

WORLD CHAMPION SABBATICALS

In the first thirty years or so of the sport when a world champion claimed to be retiring he was indeed retiring, end of story. That applied to **Farina, Fangio, Hawthorn, Brabham, Phil Hill, Graham Hill, John Surtees, Denis Hulme, Jackie Stewart**. A few, namely **Ascari, Clark** and **Rindt** did not get the benefit of retiring on their own terms, having died from accidents.

There was a bit of hiatus, from 1975 until 1978, when no world champions retired. Then, in a short three seasons, between 1979 and 1981, five of them, **Hunt, Lauda, Scheckter, Fittipaldi** and **Jones**, would drop out of the sport. In general, the reason given was they did not fancy the new breed of ground effects racing car, which supposedly removed skill from the success equation.

Since then, some of the sport's greatest champions did retire from Formula 1, then unretired, with various outcomes. Here they are.

NIKI LAUDA

After abruptly quitting **Ferrari** before the end of the 1977 season, Niki Lauda also quit his next team, **Brabham**, and did one better, shockingly retiring from Formula 1 in 1979 although relatively young, at 30 years old. Having learned to fly, Lauda got the notion that running an airline should not be all that difficult and founded Lauda Air. Despite his best efforts, it would take many years before the company took off the ground, mostly due to financial considerations. As a result it did not take too much to convince the Austrian to come back to Formula 1 in 1982, hired by **McLaren**, and he became the first world champion to unretire. He did well in his comeback, winning two races in 1982, and then his third championship in 1984, beating his teammate **Prost** by half a point. He would stick around another year, winning his final GP in 1985, at Holland. With a bit more money in his wallet, the Austrian was able to finally get his airline project working, in 1985. The company would last until 2013, by which time Lauda was no longer affiliated with it.

ALAN JONES

(Rob Neuzel)

The Australian debuted in Formula 1 in 1975, after a relatively bad start to his European career in the late 60s.

Matters improved around 1973, when Jones became one of the top Formula 3 and then Formula Atlantic drivers. In F1, after short stints at **Harry Stiller**'s privateer team, **Surtees** and **Shadow**, Jones settled in the **Williams** team, in what looked likely to be a long association in the **Chapman-Clark** mold. After a fair debut season, in 1978, by mid-1979 Williams had become the fastest car on the grid, and Jones the fastest driver. He did what was expected of him in 1980, winning the Championship and more races than anybody else. He continued to win in 1981, in fact won the first and last races of the Championship, but in hindsight Williams made a mistake having two top drivers in the team. **Carlos Reutemann** was hired for the 1980 season, and between the Belgian GP of 1980 and the Belgian GP of 1981 the Argentine had an incredible run of 15 consecutive points paying positions, which was mighty difficult at the time, as even good cars would break often and only the top 6 scored. As a result, Reutemann became the team's favored driver for the 1981 title, which certainly ruffled some feathers in the Australian's cap. At the end of 1981 Jones retired from Grand Prix racing. Driver management has never been Williams' forte, and Reutemann would last all of two races in the 1982 season, retiring as well.

Jones obviously felt he had unfinished business in F1, after all, he raced in only seven seasons, and came out of retirement in the Long Beach Grand Prix of 1983, driving for **Arrows,** after running in a non-championship race. Having qualified 12th, the Aussie retired due to driver discomfort, and at any rate, could not agree with the terms of a longer commitment. So, Jones vanished again, until coming back as one of the drivers in the new **Carl Haas** team in 1985. His time at Haas was mostly miserable, retiring from most races, generally qualifying in the last third of the grid, and not even the new **Ford Turbo** engines helped his cause. In 1986 he got four points from fourth in Austria and sixth at **Monza**, and that was the end of

Formula 1 for Alan Jones, who would still race in sports cars for many years, in Australia and Europe.

ALAIN PROST

Ferrari was in a terrible situation in the early 90s, for its last driver title dated back to 1979. After hiring **Mansell** for the 1989 season, Maranello got **Prost** to join the team for 1990. In the first season Prost did well, challenging **Senna** for the title and winning five times. The 1991 642 type was not a great car, and frustrated Prost greatly, so late in the season he referred to the car in very unflattering terms, calling it a truck. That was the last straw: Ferrari fired the 3-time champion on the spot, for linking the red car to a truck was tantamount to a crime in Italy, and surely contractually forbidden. For 1992 Prost had no option but to take a "sabbatical" which many felt was a euphemism for retirement. He could not go back to **McLaren, Williams** had **Mansell, Benetton** hedged all its bets on new hero **Schumacher**, and for obvious reasons, Ferrari would not take him back. There was simply to other suitable team for the Frenchman.

It did not take him long to come back. Williams could not agree with Mansell's financial terms for 1993 and had probably become wary of the Brit's habitual whining about the car, even when he was winning everything in sight! As Mansell was heading to the CART series in the USA, Prost was coming to Williams, a state of affairs which certainly pleased **Renault** that was not able to give the Frenchman a championship in his 3-year stint there. Prost raced well, won seven races and the title that season, and then decided to retire for good. The prospect of sharing Williams with Ayrton Senna, who half-jokingly said he would drive for free for the team in 1994, certainly had something to do with it…

NIGEL MANSELL

Nigel did not retire from Formula 1, that is true, he was simply dissed by **Williams** at the end of the 1992, who did not like the idea of paying him mountains of money to drive a car that, the boss felt, half of the field could win with. Let us call this an involuntary sabbatical. As a result Mansell went to the USA, and driving for the **Newman-Haas** team did incredibly well winning the competitive series and five races. For all intents and purposes there were no plans to return to F1 in 1994, for the same reason **Prost** sat out the 1992 season: all the good seats were taken. Then **Senna** died, and Williams was on the spot. In hindsight it appears as if Williams never trusted **Damon Hill**'s talents all that much, and **David Coulthard** was just a promising rookie, so Mansell was hired (for a mountain of money, it turned out), to do four races for his old team. Nigel did very well winning the Australian Grand Prix, and then got his name on a **Marlboro McLaren** contract for 1995. That was the initial season of cooperation between McLaren and Mercedes, but from the start things did not go well, as Mansell skipped the first two races. The cockpit was not designed to fit Mansell's expanding girth, so it had to be redesigned to fit the star driver. When he did hit the tracks, he was not that fast, giving up in Barcelona due to atrocious handling, and walking out of a lucrative contract. That was finally the end of Nigel's long F1 career.

MICHAEL SCHUMACHER

Before entering Formula 1 in 1991, Mercedes Benz supported Michael Schumacher's career in Formula 3, then in Sports Cars. It was just a matter of time before **Mercedes** would come back to Formula 1, and Michael would probably end his career driving for the team that gave him his first chances in the sport. After a race at **Jordan** and four seasons and change at Benetton, Schumacher found himself at **Ferrari**, and in 2000 had changed the venerable team's luck, by winning its first driver's title since 1979. He then won four additional straight titles, broke most records, and while still very competitive, retired from the sport as a Ferrari driver in 2006, at 38 years of age. He continued active, driving racing motorcycles, practicing all types of radical sports, but then a call came from the past: Mercedes would return as a works team in Formula One, and Schumi's services were required. The German would return to F1 in 2010, 41 years old, and would have at his disposal cars prepared by a team that had won the previous year's championship, albeit under a different name. In retrospective, it was a mistake. While most people were expecting Michael do to a **Niki Lauda-return**, quite the opposite happened, it was just an improved **Alan Jones-return**. In his 3-year return Schumacher was obviously slower than teammate **Nico Rosberg**, and had a single podium out of 37 races, a few laps on the lead, a few exceptional qualifying efforts, and was never a factor in any of the seasons. He finally retired for good in 2012, a Mercedes driver, but his thirst for adrenaline would get the best of him.

KIMI RÄIKKÖNEN

The Finn did it his way, like **Frank Sinatra**. He came into Formula 1 very young, having an amazingly short car racing curriculum, and within two seasons had gotten his first win and runner-up position in a championship driving a somewhat inferior car. Then he won the 2007 title for **Ferrari** under extraordinary circumstances, winning the first GP of the season, then most of the races in the last quarter of the year. He would stay at **Ferrari** an additional two seasons, but by the end of 2009 Ferrari bought him out of his contract, essentially paying him not to drive, to make room for **Fernando Alonso**. After the Alonso - **Hamilton** fiasco of 2007, perhaps it was wise not to have two world champions running at Ferrari at the same time. Another Finn retiring early? Well, no one said Kimi was retiring. He did rallying, ran NASCAR trucks and stockers, and had some fun in his two years out of the sport, and had money to burn. In 2012, he was contracted by **Lotus F1,** which was really **Renault** by another name. There was a performance clause to his contract: he would be paid a 5-million euro fixed retainer, plus 50,000 Euros for each point scored. I guess they did not trust he would get all that many points, but that is exactly what he did, 207 of them, 3rd place in the championship plus a race win. To avoid going bankrupt they reduced his fixed retainer to 2 million in 2013, but that did not matter: he got 183 points and lots of money for his efforts, plus another win! In a sense Lotus F1 was kind of happy to let Kimi go back to Ferrari, where he would partner…Fernando Alonso. It turned out that Kimi was not a difficult teammate, worked well with Fernando, then with **Vettel** who replaced the Spaniard in 2015. Kimi's second stint at Ferrari was not as successful as the first one, but he became the most popular Formula 1 driver, despite his unsmiling face and curt answers, or perhaps because of it, and would still get another win on the books, in the US GP of 2018. He also finished 3rd in that final Ferrari season, after taking 4th places in 2015 and 2017. Not quite ready to retire, Kimi spent three seasons driving for the **Alfa Romeo** team, where life at the bottom end of the grid proved increasingly frustrating. He finally retired from Formula 1 at the end of the 2021 season but will soon be trying NASCAR racing again.

Fernando Alonso never quite announced a retirement or even sabbatical from Formula 1, but he stayed out in the 2019 and 2020 seasons. Since 2017 he had been trying to match **Graham Hill's** record, attempting to win both the Indy 500 with McLaren, his F1 team at the time, and the 24 Hours of Le Mans with **Toyota**, but succeeded only in the French classic, not the American one. He also raced in the Dakar Rally. In 2021 he rejoined Formula 1, hired by **Alpine**.

Then there are the "almost returns" by world champions.

Emerson Fittipaldi was still a young man when he retired from Formula 1 in the end of the 1980 season. After five years trying to make the **Fittipaldi team** a member of the F1 establishment, Emerson was tired. He had no choice: it was either retiring or continue driving for his own team in 1981. In that juncture he was no longer considered a viable choice for top teams, so he would likely get an offer from a midfield team, if at all. It turned out the situation at the Fittipaldi team got worse in 1981, having lost Skol beer sponsorship, thus Emerson retired.

Four years later, in 1984, there were rumors of an Emerson return. He actually tested the **Spirit** Formula 1 car, designed by **Gordon Coppuck**, the designer of the **McLaren M23** that brought him the 1974 championship, but sensibly, refused a race seat. That was a smart decision, for the move would probably blot his copybook further, as Spirit would not last beyond the 1985 season and was never competitive. In the end, he did unretire, starting a successful career in Indy Cars that lasted until 1996.

Then there was **Mika Häkkinen**. One of the fastest drivers of his generation, Mika only enjoyed four seasons with the possibility of winning the championship and won two crowns, not a bad ratio. In 2001, even younger than Emerson, he decided to take a sabbatical for an undisclosed period, and McLaren guaranteed he would have a seat back whenever he wanted. Ultimately he never came back, and by 2003 another fast Finn was racing at McLaren, **Räikkonen**. However, Mika did resume his racing career, driving for **Mercedes** in the DTM for three seasons, finally calling it a day at the end of 2007.

THE INDY 500 ANGLE

Many Formula 1 books simply avoid the fact that the Indy 500 was a part of the World Championship of Drivers between 1950 and 1960. That is understandable: the cars that raced at Indy were substantially different from the F1 cars, the race took place in an oval track, and the participants in one discipline largely ignored the other. However, the fact is that dozens of American drivers appear in the final Championship results for those years, after all, the race was part of the championship. When you visit some statistics sites, you will note many American drivers and also constructors: do not fear, you are not going crazy and yes, they raced only at Indy, with a single exception.

Bill Vukovich was the only Indy driver to win the 500 twice in the 1950-60 years, while the race was part of the World Championship. He also led 485 laps and participated in the race five times. (Unattributed photo)

The **Indy 500** was the longest race in most seasons, the 1951 edition clocking 3 hours 57 minutes and 38 seconds, and in most seasons, it was the race with the largest number of starters, always 33. Its inclusion obviously skews the numbers in favor of the Americans.

In the 1950 to 1960 period, no less than 220 **Kurtis-Kraft** roadsters started the Indy 500, while a large number did not qualify. That does not mean that other chassis did not participate. In fact, I counted 54 other chassis manufacturers attempting a place under the sun in this period: **Adams, Bardazon, Bromme, Cantarano, Christensen, Christy, Clemons, Cornis, Dedit, Del Roy, Dunn, Elder, Eppery, Ewing, Gdula, Gerhardt, Hall, Hillegass, Johnson, Koehnle, Kuptec, Kuzma, Langley, Lesosvky, Marchese, Meskowski, Meyer, Miller, Moore, Nichels, Olson, Pankratz, Pawl, Phillips, R. Miller, Rassey, Rounds Rocket, Schroeder, Scopa, Sherman, Shilala, Silnes, Snowberger, Stevens Sutton, Szalai, Templeton, Trevis, Turner, Voelker, Watson, Watts, Weidel, Wetteroth**. Only a few survived beyond the decade, mostly becoming providers of Midget and Sprint cars after the 70s.

Adding insult to injury, most of the results you see in the record books and online identify the cars under the names of sponsors and car owners, rather than chassis producers. The vast majority of the cars used the **Offenhauser** engine, but even **Cadillac** and **De Soto** engines were unsuccessfully tried.

If someone asks, you can say the great Indy champion **A.J. Foyt** has raced in the World Championship for Drivers three times, after all he began his career in the late 50s. He missed getting a World Championship win by one year, for he won his first Indy 500 in 1961.

Rodger Ward entered a midget car (not an Indy car) in the 1959 U.S. Grand Prix, the same year he won Indy for the first time. **Alberto Ascari** was the only F1 driver to actually start the race in this period, while **Giuseppe Farina** tried a few times, and **Juan Manuel Fangio** once (in 1958), both without success.

The first Indy car (USAC at the time) champion to make it in F1 was **Mario Andretti**, who raced in F1 from 1968 to 1982, having won the World Championship in 1978. The Americans that gravitated to F1 were generally sports car drivers, such as **Gurney, Ginther, Hill**, etc. Former CART champion (1995) **Jacques Villeneuve** also became a Formula 1 Champion in 1997, and **Juan Pablo Montoya** came up short.

There were several attempts to equalize Formula 1 and Indy racers, including high level meetings held before the 1970 season which called for a standard 4-liter engine for Formula 1, Indy and World Sports Cars. These discussions did not progress, and the disciplines have stayed apart to this day, even though driver exchange between the categories increased since the 80s and European design practice has been adopted since the mid-60s.

WHEN FORMULA ONE IS NOT REALLY FORMULA ONE

In other parts of the book you have found that cars from a number of other categories took part in World Championship (Formula One) races: Formula 2, Formula 3 with enlarged engines, Midget, Champ Cars, even sports cars. Now, let us consider cars that were referred as F1, but were really not Formula One cars.

Do not be confused when you do internet searches and find out a few "Formula 1" championships occurred worldwide back in the old days. These were not run to international F1 regulations, and often were not truly F1 cars.

This is the Campo-Chevrolet, a Mecanica Argentina F1 (Alejandro de Brito)

Take, for instance, MAF1 (Mecanica Argentina Formula 1). This single seater category ran from the 60s until 1979, and races for this category were held in Argentina, Uruguay and Paraguay. The participants were single seaters built in Argentina by the likes of **Berta, Pianetto, Campo, Bravi, Sotro** and a few dozen others, and equipped with Argentine built engines with capacity of as much as 4 liters (**Dodge, Tornado, Chevrolet** and **Ford**). The regulations also allowed cars with smaller sized engines such as **Peugeot**, and lower weight, but these were not competitive. The cars were much slower and less powerful than international Formula 1. Other categories named Formula 1 raced in Argentina after the demise of MAF1 in 1979, and needless to say, these had no connection whatsoever with international Formula One.

There was an even slower Formula 1, a category that ran in the Soviet Union in the 60s to mid-70s. This was the top single seater category in the region but can only be called F1 with a lot of poetic license. Soviet media at the time sporadically hinted that cars such as the **MADI,** would be entered in the World Championship, but this was just propaganda at its worst.

The MADI-01 in 1974 driven by Stanislav Gess-de-Kalve, its designer (Estonian Motorsport Archive)

Sometimes the Western press would reproduce such fakenews, for the subject generated interesting press. Many of the Soviet F1 cars had production based engines (such as Moskvich) that could produce about 120 HP, and, at best, there may have been one or another that generated 300 HP, but that is questionable and unlikely. In 1977 the Soviet F1 was replaced by Formula Easter (Vostok). These cars ran in Russia, Belarus, Ukraine, Estonia and Lithuania.

Then there was South African Formula 1. The first edition of the South African Grand Prix valid for the World Championship, held in 1962, had a few unusual entries, which would race only in that round of the championship. These were single seaters produced in South Africa, such as the **LDS**, and equipped with production-based engines such as **Alfa Romeo** and **Ford**, that stood no chance against pure racing engines such as **Climax, BRM** and **Ferrari**. With the onset of 3-liter regulations, these "specials" disappeared, although LDS entered cars until 1968, by then with **Repco** engines. The South African Formula 1 Championship per se would run until 1975, but before you start telling your friends that tons of F1 cars raced in local South African races you should know that fields were padded with Formula 2 and Formula 5000 cars, and even so, grids were never very large, and F1 cars very few. The European press generally made it look more prominent than it was.

The Australian Formula 1 Championship, on the other hand, was not ran to International Formula 1 regulations for the most part. The Aussies and New Zealanders adopted the 2.5-liter engine regulation for the rest of the 60s, which became known as Formula Tasman, and starting in 1971, Australia adopted Formula 5000, which were single seaters with up to 5-liter engines. The Championship continued to be called Australian Formula 1 Championship until the early 80s.

True formula 1 cars were accepted in the Australian Rothmans' International Series (which replaced the Tasman Series in Australia) in 1979, in an attempt to attract foreign entries. An **Ensign** and a **Wolf** were entered in the series, and **David Kennedy** won at Surfer's Paradise, driving the Wolf.

Until the early 80s non-championship Formula 1 races (which will be covered in a different volume) were held in Europe and the Americas, which often outnumbered the Championship races. Most of these races included Formula 2, odd sports cars and later on, Formula 5000 cars, so they were actually Formula Libre races.

With the demise of the European Formula 5000 Championship in 1975, a Group 8 Championship was established with races mostly in Britain, which accepted Formula 1, Formula 5000, Formula 2 and even Formula Atlantic cars. This eventually became the Aurora Championship, by which time only Formula 1 and Formula 2 cars were accepted.

Formula 1 cars also raced in a few categories and championships, for instance, 3.0-liter F1 cars were entered in the US Formula A/Formula 5000 Championship until 1972 and even won races in that championship. In the early 80s, F1 cars were raced in the revived Can-Am, which was established in 1977 as a replacement for Formula 5000. The category comprised of single seaters with sports car bodies and engines up to 5 liters. These F1 cars ran with slightly enlarged **Cosworth** DFV engines to 3.3 liters, so they were not "proper" F1 cars, which, at the time, had a limit of 3 liters for normally aspirated engines. A F1 **Ensign** thus managed to win a race under these conditions.

Fendered Formula 1 cars such as **Footwork, Fondmetal, Jordan** and **Minardi** also appeared in the Interserie championship in the mid-90s, racing against cars such as **Porsche 962** and other Group C sports cars, fendered Indycars and older Can Am racers. These fendered Formula 1 cars won races as well.

Then there is always exaggeration. An article from Kenya in the 70s refers to Formula 1 cars being raced in the local Nakuru racetrack, but these were likely old F3 (50s vintage) or Formula Junior cars, or an older F1 chassis mated to a production engine of some sort (most likely a **Ford**).

Older Formula 1 cars have also found an afterlife in many corners of the world. A few **Maserati** 250Fs were raced in many countries after F1 duty, such as Argentina, Brazil, Australia, New Zealand, and the US. In South America these cars were equipped with American 6- and 8-cylinder engines which were less expensive to run. Other Grand Prix chassis such as **Ferrari, Lancia, Talbot** and even 1930s **Bugattis** also ran in these races. These hybrid categories disappeared around the mid-60s. One of the drivers that raced a Maserati 250F (with the original engine) in local racing was **Chris Amon** in his native New Zealand.

Formula 1 cars have also been raced in hill climbs in many countries. Most famously, **Jean-Pierre Beltoise** won a round of the European Hillclimb championship in 1970, at Ollon-Villars, driving a Formula 1 **Matra-Simca**. F1 chassis equipped with DFVs or other engines, were also raced in the British Hillclimb Championship. The **Token**, a very unsuccessful F1 design, came close to winning British Sprint Championship events of the late 70s.

A demonstration Formula 1 "race" was held in Japan, in 1974, in preparation for the 1976 Grand Prix. A small field participated, including **Fittipaldi, Reutemann** and **Peterson**, but it seems that the Swede was the only one interested in running fast and won the "event".

Then there is historic racing, which involves F1 cars from different periods and regulations, some very old and preceding the World Championship age. These cars are generally not raced "hard", for some of them are unique and worth millions of dollars, but every once in a while, one crashes. **Charles Leclerc** crashed a **Ferrari 312B3** at Monaco, 2022, during a demonstration run.

A WHOLE BUNCH OF JOHNS

It is no secret that baby names fall in and out of favor every few years. Although there is some logic to it (for instance, the rare name Alanis has become more common recently, due to a famous Canadian singer who rose to fame in the 90s), in other cases it is not possible to pinpoint the widespread use of certain names at given periods. Take for instance, the name John (and Juan, Jean, Johann, Giovanni).

Sure, John is a popular name in the Western world, as one of Jesus' apostles was name John. But in the early days of Formula 1 the almost omnipresence of Johns at the top echelon of the sport borders on the insane.

While the first F1 champion was named Giuseppe (Joseph in Italian), the very second champion was a John: **Juan Manuel Fangio**. Then came **Alberto Ascari**, who won two straight titles, and Fangio came full strength and won four straight titles.

The next on the list was **Mike Hawthorn**. Really? It so happens that Hawthorn's full name was **JOHN Michael Hawthorn**. That would be curious indeed, except that the guy who won the next two titles, who was normally known as Jack Brabham, was really named **JOHN Arthur Brabham.**

So there you have it, eight F1 championships from 1950 to 1960 were won by guys named John. Plus American **Johnnie Parsons** won the 1950 Indy 500.

Things seemed to settle down, as a Phil, a Norman (Graham) and a James won titles in succession. Then came another **JOHN, Surtees**, who won in 1964. In 1965 another John appeared on the scene, looking like a good bet for future world champion. Known as Jackie, his name was **JOHN Young Stewart**! And he would fulfill the forecasts, by winning the 1969, 1971 and 1973 world titles.

So there you have it, between 1950 to 1973 twelve F1 championships were won by drivers named John, quite a remarkable fact. Since then no further John won the F1 championship, so I suppose that starting in 1939, Jackie Stewart's date of birth, the name John fell out of favor. At least among parents of future world champions.

But Johns continued to race and win in F1. Three French Jean-Pierres raced from the 1960s to the early 1980s, and two of them, **Beltoise** and **Jabouille**, won races, plus a Northern Irish driver, **John Watson**, won races starting in 1976 until 1983. In the 1990s it was the turn of **Johnnie Herbert** to win three Grand Prix for Benetton and Stewart, while **Jean Alesi** won his single GP in 1995.

So far, the victory swansong for the brotherhood of the Johns was the 2000s. Italian **Giancarlo Fisichella** won three Grand Prix, while Colombian **Juan Pablo Montoya** looked like a possible future John champion, who finished 3rd in the championship twice (2002 and 2003) and won seven races. Since then, the Johns have calmed down, and in fact Johns/Jeans/Juans/Giovanni have been mostly off the grid since the Montoya days. Not even the last Jack to race in F1, Aitken in 2020, is a John: his name is really Jack!

In more recent times, since the early 2000s, the name Sebastien/Sebastian has become one of the most common at racing's top level, another generational issue, I suppose. Not only did **Sebastian Vettel** win four F1 titles, but two other Sebastiens, Buemi and Bourdais, failed in F1 but did extremely well at Le Mans and Indycars. Plus for most of the last 20 years the world rally champion has been a Frenchman named Sebastien, either Loeb and Ogier.

THE GRAND PRIX OF MONACO: GLAMOUR, EXCITEMENT AND BOREDOM

Every so often rumors fly that the Grand Prix of Monaco will be cancelled in short order. At present, the organizers are pressed by Liberty Media to sign a new contract, for allegedly Monaco pays a relative pittance for the right to hold the race compared to other GPs. Whether one likes or not the Monaco race, the fact is that arguably it is the only race in the F1 calendar that could stand on its own, much like the 24 Hours of Le Mans and the Indy 500 do in Sports Car and Indycars. Attempts to bring as much glamour to F1 outside of Monaco have all failed: Long Beach, Yas Marina, Valencia, and now the life-size boat dioramas of the Miami Grand Prix all fell short of the attraction of the Principality, whose population is largely comprised of wealthy individuals, including current and former Grand Prix drivers. Other present and past street races, such as Azerbaijan and Detroit cannot even dream of aspiring to Monaco status. Liberty Media now plans to make Las Vegas __THE EVENT__ of the season, but besides hefty (read unaffordable) ticket prices, it remains to be seen whether this will come to fruition. For sure there will be tons of celebs who may think that Formula 1 is a hemorrhoid medicine but will get freebies to walk the grid, look interested and give Martin Brundle the cold shoulder. The difference is that Liberty Media is organizing the Vegas race, so it will have all risks, but also stands to make all income.

Held since 1929, the Monaco Grand Prix occurs in one of the smallest countries in the world, for Monaco has an area of 0.78 mi^2 (a little less than 2 km^2) in a spectacular setting, between mountains and the Mediterranean Sea, in the French Riviera and only about 10 miles from Northern Italy.

The country became independent in 1861, and currently, about 7 out of 10 of its 38,000 residents are allegedly millionaires, and their income is not taxable. Before you think of moving there to hobnob with the rich and famous, beware, even renting a small place in Monaco is extremely expensive, five-figure expensive, and an espresso will make you 10 bucks poorer at normal times. The GDP per capita is roughly US$ 235,000 (in comparison, the USA's is US$ 70,248), which helps explain the number of yachts parked in the marinas come race day.

The race was created by Antony Noghes, who also conceived the equally traditional Monte Carlo Rally in 1911. Noghes's genius was seeing the promotional value that a top racing event (or a sports event, for that matter) could bring to the principality as a tourist destination. Noghes also allegedly suggested the use of checkered flags to end races, so his DNA is found in almost every car race in the world.

Lovers of Formula 1 have complained of the long processional races at the Principality, brought about by the tight confines of the narrow track and short straights, which makes overtaking next to impossible. The Fairmont hairpin, for example, is taken at merely 30 miles per hour. Nevertheless, the Grand Prix of Monaco has been the site of excitement over the course of the years, caused by accidents, proximity to the sea, lapses in concentration, rain and retirements in late stages of the race.

In 1950 Monaco was the second ever World Championship race and witnessed a huge pile-up that eliminated nine cars before completion of the first lap. A huge wave crashed onto the sea wall, bringing a lot of water onto the track. First race winner and eventual champion Giuseppe Farina spun and crashed his Alfa Romeo, and his sideways car and wet track caused the cars of **Schell, Rol, Harrison, Trintignant, De Graffenried, Manzon, Rosier, Fagioli** to retire before completing a single lap. Thus **Juan**

Manuel Fangio won his first Formula 1 Grand Prix at Monaco, and he is not alone among World Champions. **Jack Brabham** and **Denis Hulme** also won their first Grand Prix at Monaco, and non-champions **Trintignant, Beltoise, Panis, Depailler, Patrese** and **Trulli** also won for the first time there.

The messy beginning of the 1950 race, which was not interrupted. (Alejandro de Brito)

Proximity to the sea also caused a problem in the 1955 race. **Alberto Ascari**, who had won the 1952 and 1953 titles for **Ferrari**, was driving a **Lancia** and looked set to win the 1955 event as leader **Stirling Moss' Mercedes** blew the engine in the late stage of the race. This caused Ascari to lose concentration, so coming out of the famous tunnel, Alberto lost control of the car, which crashed through the barriers into the Mediterranean. Luckily Alberto knew how to swim, but a few days later he would die testing a car in Monza.

A slow circuit, Monaco has favored driver-car combinations that may not have been the fastest of the day. Such is the case of the 1961 race, won by **Stirling Moss** in a down-on-power **Lotus-Climax**. That season **Ferrari** had the upper hand in terms of horsepower, but in Monaco Moss managed to prevail with his fine driving skills, even though

Ginther was closing the gap to him in the final laps. Most famously, a side body panel fell from the Lotus, so that the public could watch Moss' movements inside the cockpit.

Graham Hill was the first Monaco king and won his first race there in 1963 driving a **BRM**. He would also win 1964, 1965, 1968 and 1969, so that five of his 14 wins, including his last one, took place in the Principality. A driver of a different mold, the extremely fast **Ayrton Senna**, would overtake Graham winning Monaco a total of six times, including five straight wins from 1989 to 1993. Senna came close to ratcheting up seven straight victories, but a rare lapse in concentration in the 1988 race led to his retirement. That was the year of Senna's first championship, and in qualifying he was almost 1.5 seconds faster than second placed **Alain Prost**, driving what many consider the most perfect Monaco lap ever, described by Ayrton as a religious experience.

Senna also came close to winning his first Grand Prix there, in 1984, a result that turned out to decide the champion in the closest battle ever. Under heavy rain Senna, who started 13th, was coming closer to the leader Prost as many of the fastest qualifiers such as **Arnoux** and **Mansell** hit trouble at the track. Driving the sluggish **Toleman-Hart** Senna was a little over 7 seconds behind Prost and catching up fast when Clerk of the Course **Jacky Ickx** red flagged the event on security concerns. Third placed **Stefan Bellof** was also catching both Prost and Senna, and apparently would pass them both sooner or later all things being equal, but his team **Tyrrell** would be disqualified from the Championship, so the result would not stand anyhow. In Brazil some overly excited fans (and press) felt that the race was given to Prost in detriment of Senna, but the fact is that Prost lost the Championship because he scored only 4.5 points from his half-win, whereas had he finished second in the full event he would score 6 and possibly win the Championship, ultimately won by **Niki Lauda** by a mere 0.5 point. Lauda did not score at all that day.

Rain in 1972 resulted in the longest F1 race of the modern era, which lasted 2h26m55s without interruption and the result was popular, for the winner was **Jean-Pierre Beltoise** driving a **BRM.** That turned out to be Beltoise's single F1 win, BRM's last and Marlboro's first win as a major F1 sponsor. Many of the fastest drivers of the day (**Emerson Fittipaldi, Jacky Ickx, Jackie Stewart**) had problems keeping their cars on track in the downpour, but amazingly only six of the 25 starters had accidents.

The same cannot be said of another Monaco Grand Prix won by a Frenchman who also had his only Formula 1 win on the streets of the Principality. **Olivier Panis** won the 1996 race starting in 14th place, and in the end was the first of three cars to actually take the checkered flag, while a fourth car was in the pits. The crash fest began in warm-up, with **Montermini** in the **Forti**. In total 13 cars collided, had accidents or spun, and only four cars retired with mechanical problems. Before the first lap was completed five cars had been eliminated, so **Ligier** won a GP for the final time before changing ownership and name in 1997.

Another driver who won his single race at Monaco was **Jarno Trulli**, a fast Italian who usually started in the top 10 even with bad cars, had four poles and led a total of 163 laps in his career. Somehow, Jarno had a problem keeping his rhythm for entire GP lengths, and usually a queue would form behind him as the races evolved, "endearingly" referred as the Trulli Train. That obviously did not make him popular with his colleagues who were held up and fans of other drivers, notwithstanding Jarno remained in Formula 1 until 2011. In the 2004 edition, driving for Renault, Trulli scored pole and led most of the laps (72), ending half a second in front of **Jenson Button**, who was driving the **BAR-Honda**.

A most unusual Monaco Grand Prix was the 1982 race. **Rene Arnoux** started the race from pole but spun out of the lead. **Prost,** his **Renault** teammate, took over the lead until three laps to the end of the Grand Prix. Alain made a rare mistake and crashed, so **Riccardo Patrese** took over. The **Brabham** driver spun on some oil, and his car stalled on track. He was passed by **Didier Pironi**, whose car had electrical problems and stopped in the tunnel during the very final lap. Didier was about to be passed by **Andrea de Cesaris**, but the Italian ran out of fuel in his thirsty **Alfa Romeo**. It appeared as if **Derek Daly** would display some Irish luck, but he also had an accident which damaged his gearbox. Patrese did get his car restarted rolling it downhill and managed to win his first race. Both second and third place cars (Pironi and De Cesaris) did not take the checkered flag, and in total, only four other cars did.

In 1970 **Jack Brabham** lost the Monaco Grand Prix in the very final lap of the race. Jack had been racing in F1 since 1955, and this was to be his last season. He started the year well, winning the South African Grand Prix, and looked as racy as he did in the 1960 and 1966 seasons, when he won his second and third titles. At Monaco he started fourth, and **Stewart** readily jumped in the lead until his engine developed problems, and he was overtaken by the Australian on the 27th lap. Brabham seemed on the way to a clear victory, only that **Jochen Rindt**, who had started in 8th place, began to close the gap. On the final corner of the last lap Brabham made a mistake, locked the wheels and the car spun, leaving the road open for Rindt who won his first race of the season. Jack still managed to get the car going, finishing in 2nd place and leading the championship again. Amazingly, Brabham would also lose the 1970 British Grand Prix to Rindt on the final lap, for a different reason.

These are just some of the dramatic moments of the history of the Monaco Grand Prix. While it is true that most of the time racing has been processional, and downright boring, the grand event seems to spring about surprises, and even when it does not, seeing lucky people watching the race from balconies at Montecarlo sipping champagne strikes a better note than seeing a similar scene in Baku, drinking who knows what.

WHERE CHAMPIONS RACED AT ONE TIME

World champions have raced for different teams and constructors during their careers, in fact, quite a few dozen of them. Not surprisingly, **McLaren, Ferrari, Lotus** and **Williams** head the list, even though more champions have raced for McLaren than Ferrari. In some cases, these were participations, with DNQ or non-starts, like **Fangio**'s and **Farina**'s attempts to qualify at the **Indy 500** driving **Kurtis-Krafts**. (*) indicates the driver won race(s) for the marque, and (+) indicates driver won championship(s) for the marque. One will notice that all champions who raced for Ferrari won at least one race, which is not the case with McLaren, Williams and Lotus. The only one who raced his entire career for a single constructor was **Jim Clark**.

MCLAREN

E. Fittipaldi (*) (+)
A. Senna (*) (+)
A. Prost (*) (+)
N. Lauda (*) (+)
L. Hamilton (*) (+)
M. Häkkinen (*) (+)
J. Hunt (*) (+)
D. Hulme (*)
F. Alonso (*)
J. Button (*)
K. Räikkönen (*)
J. Scheckter
N. Piquet
N. Mansell
J. Surtees
K. Rosberg
P. Hill

FERRARI

A. Ascari (*) (+)
J.M. Fangio (*) (+)
M. Hawthorn (*) (+)
P. Hill (*) (+)
J. Surtees (*) (+)
N. Lauda (*) (+)
M. Schumacher (*) (+)
K. Räikkönen (*) (+)
J. Scheckter (*) (+)
A. Prost (*)
N. Mansell (*)
F. Alonso (*)
M. Andretti (*)
S. Vettel (*)
G. Farina (*)

LOTUS

J. Clark (*) (+)
G. Hill (*) (+)

M. Andretti (*) (+)
J. Rindt (*) (+)
E. Fittipaldi (*) (+)
A. Senna (*)
K. Räikkönen (*)
M. Häkkinen
J. Surtees
N. Mansell
P. Hill
J. Brabham
N. Piquet

WILLIAMS

N. Mansell (*) (+)
A. Prost (*) (+)
D. Hill (*) (+)
A. Jones (*) (+)
K. Rosberg (*) (+)
J. Villeneuve (*) (+)
N. Piquet (*) (+)
N. Rosberg
J. Button
M. Andretti
A. Senna

BRABHAM

N. Piquet (*) (+)
J. Brabham (*) (+)
D. Hulme (*) (+)
N. Lauda (*)
J. Rindt
G. Hill
D. Hill

COOPER

J. Brabham (*) (+)

J. Surtees (*)
J. Rindt
P. Hill
M. Hawthorn

MASERATI

J.M. Fangio (*) (+)
M. Hawthorn
J. Brabham
A. Ascari
P. Hill

MERCEDES

J. M. Fangio (*) (+)
L. Hamilton (*) (+)
N. Rosberg (*) (+)
M. Schumacher

ALFA ROMEO

J.M. Fangio (*) (+)
G. Farina (*) (+)
M. Andretti
K. Räikkönen

MARCH

J. Stewart (*)
M. Andretti
N. Lauda
J. Hunt

BRM

G. Hill (*) (+)
J. Stewart (*)
N. Lauda

M. Hawthorn
J. Surtees

BENETTON

M. Schumacher (*) (+)
N. Piquet (*)
J. Button

LOLA

J. Surtees
G. Hill
A. Jones

WOLF

J. Scheckter (*)
J. Hunt
K. Rosberg

TYRRELL

J. Stewart (*) (+)
J. Scheckter (*)

RED BULL

S. Vettel (*) (+)
M. Vertsappen (*) (+)

TORO ROSSO

S. Vettel (*)
M. Verstappen

FITTIPALDI

E. Fittipaldi

K. Rosberg

HONDA

J. Surtees (*)
J. Button (*)

SURTEES

J. Surtees
A. Jones

BAR

J. Villeneuve
J. Button

SAUBER

J. Villeneuve
K. Räikkönen

SHADOW

A. Jones (*)
G. Hill

HESKETH

J. Hunt (*)
A. Jones

RENAULT

F. Alonso (*) (+)
A. Prost (*)

HILL

G. Hill
A. Jones

JORDAN

D. Hill (*)
M. Schumacher

KURTIS-KRAFT

G. Farina
J.M. Fangio

MATRA

J. Stewart (*) (+)

BRAWN

J. Button (*) (+)

TOLEMAN

A. Senna

VANWALL

M. Hawthorn

EAGLE

P. Hill

ALPINE

F. Alonso

BMW-SAUBER

S. Vettel
J. Villeneuve

LANCIA

A. Ascari
G. Farina

ARROWS

D. Hill
A. Jones

PORSCHE

P. Hill

ENSIGN

N. Piquet

ASTON MARTIN

S. Vettel

PARNELLI

M. Andretti

MINARDI

F. Alonso

THEODORE

K. Rosberg

ATS (ITALIAN)

P. Hill

ATS (GERMAN)

K. Rosberg

BIBLIOGRAPHY

ACERBI, LEONARDO. **FERRARI ALL THE CARS**. Giorgio Nada Editore, 2008
ALDER, TREVOR & BARTON, ALAN. **FORMULA 1 WORLD CHAMPIONSHIP 1965-1967**, Transport Source Books, 197
ASH, DAVID. **GRAND PRIX ALMANAC**. Automobile Almanac Lt, 1974
AUTOCOURSE (several editions), Hazleton Securities Ltd
AUTO HEBDO, **LES 73 PILOTES FRANCAIS EN F1 1950-2018**, Auto Hebdo, 2018.
AUTODROM MOTORSPORTDOKUMENTATION (6, 7), Axel Morenno, 1974, 1975
BAMSEY, IAN. **AUTOMOBILE SPORT**. Tenorhart Ltd (several editions)
BARBIERI & VARISO. **400 CAVALLI NELLA SCHIENA**. Longanesi & Co, 1969
BEUSQUET, PATRICE, **CHARADE LE PLUS BEAU CIRCUIT DU MONDE**. EDITIONS DU PALMIER, 2003
BOWER, TOM. **NO ANGEL. THE SECRET LIFE OF BERNIE ECCLESTONE**. Faber & Faber, 2011.
BRUSINI, ROMANO. **IL COWBOY DELLE CORSE**, Brusini, 2011
CIMAROSTI, ADRIANO. **THE COMPLETE HISTORY OF GRAND PRIX MOTOR RACING**. Aurum, 1990
CUTTER, ROBERT; FENDELL, BOB. **ENCYCLOPEDIA OF AUTO RACING GREATS**. Prentice-Hall, 1973
DARMENDRAIL, PIERRE. **LE GRAND PRIX DE PAU (1899-1960)**. La Librairie du Collectionneur, 1992
DE PAULA, CARLOS. **MOTOR RACING IN THE 70S**, De Paula Publishing, 2020
DE PAULA, CARLOS, **RACING CAR CONSTRUCTORS OF THE 70s,** De Paula Publishing, 2021
DIAZ, MARIA JESUS, Editor. **EL AUTOMOVIL EN ESPANA – ATLAS ILUSTRADO**. Susaeta
DOMINIQUE, VINCENT. **MATRA TOUTE L'HISTOIRE, TOUTES LES COURSES**. L'Autodrome Edition, 2017
FIA ANNUAIRE DU SPORT AUTOMOBILE 1973, FIA, 1973
GEORGANO, G.N. **THE ENCYCLOPEDIA OF MOTOR SPORT**. Viking Press, 1971
GERAERD, BERNARD. **LA DYNASTIE PILETTE**, Appach Prodution & Publishing
GILL, BARRIE. **INTERNATIONAL MOTOR RACING, 1977.** Two Continents Publishing, 1977.
GILL, BARRIE. **JOHN PLAYER MOTORSPORT YEARBOOK**, 1972. Queen Anne Press, 1972
GILL, BARRIE. **MOTOR SPORT YEARBOOK**, 1974. Collier Books, 1974.
GRIFFITHS, TREVOR. **GRAND PRIX THE COMPLETE GUIDE**, Bloomsbury, 1997
GUBA, EDDIE. **MOTOR SPORT ANNUAL 5**, Greenville, 1972.
HAMILTON, MAURICE. **GRAND PRIX CIRCUITS**, Collins, 2015.
HIGHAM, PETER. **WORLD ENCYCLOPEDIA OF RACING DRIVERS**. Haynes, 2013.
HODGES, DAVID. **A-Z OF FORMULA RACING CARS 1945-1990**. Bay View Books, 1990
HODGES, DAVID; BURGESS-WISE, DAVID; DAVENPORT, JOHN; HARDING, ANTHONY. **THE GUINESS BOOK OF CAR FACTS & FEATS,** Guinness Publishing, 1994
HONDA, JOE. **TYRREL P-34 1976**, MFH, 2011
JAMIESON, SUSAN; TUTTHILL, PETER. **WOMEN IN MOTORSPORT FROM 1945**, Jaker/BWRDC, 2003
JENKINSON, DENIS. **A STORY OF FORMULA 1**. Greenville, 1960
JONES, BRUCE. **THE ENCYCLOPEDIA OF FORMULA 1**, Carlton, 1998
JULIAN, JOHN. 1967. **CHRIS AMON, SCUDERIA FERRARI AND A YEAR OF LIVING DANGEROUSLY**. David Bull Publishing, 2013
KETTLEWELL, MIKE. **25 YEARS OF BRANDS HATCH CAR RACING**. Brands Hatch Racing Circuit, 1975

L'ANNEE AUTOMOBILE, 1978/79, L'Edita Lausanne, 1979
L'ANNEE AUTOMOBILE, 1979/80, L'Edita Lausanne, 1980
LANG, MIKE. **GRAND PRIX! RACE-BY-RACE ACCOUNT OF FORMULA 1 CHAMPIONSHIP MOTOR RACING** (Volumes 1, 2, 3). Haynes.
LECESNE, ENGUERRAND. **CIRCUIT DE ROUEN LES ESSARTS**, ETAI
LEME, REGINALDO **AUTO MOTOR ESPORTE YEARBOOKS**, Auto Motor Editora (several Editions)
NYE, DOUG. **HISTORY OF THE GRAND PRIX CAR**, 1945-65, Hazleton, 1993
O`LEARY, MIKE. **MARIO ANDRETTI – THE COMPLETE RECORD**. MBI Publishing, 2002.
PRITCHARD, ANTHONY. **THE MOTOR RACING YEAR**, W. W. Norton, 1971, 1972, 1973
PRITCHARD, ANTHONY; DAVEY, KEITH **THE ENCYCLOPEDIA OF MOTOR RACING**, 2nd Edition. Robert Hale, 1973
PRITCHARD, ANTHONY, **THE MASERATI 250F**, Aston Publications, 1985
REDMAN, BRIAN with MULLEN, JIM. **BRIAN REDMAN: DARING DRIVERS, DEADLY TRACKS**. Evro Publishing, 2016
RIBEIRO, ALEX DIAS. **MAIS QUE VENCEDOR**. Alex Publicacoes, 1981.
SANTOS, FRANCISCO. **FORMULA ANUARIO**. Edipromo (several editions)
SCHIMPF, ECKHARD. **STUCK DIE RENFAHRERDYNASTIE**. Delius Klasing,
SCHIEGELMILCH, RAINER AND LEHBRINK, HARMUT. **GRAND PRIX DE MONACO**. Koneman, 1998
SHELDON, PAUL; RABAGLIATI, DUNCAN. **FORMULA 1 REGISTER – FORMULA 5000 FACT BOOK – (1973-1977),** St. Leonard's Press
SHELDON, PAUL; RABAGLIATI, DUNCAN; DE LA GORCE, YVES. **FORMULA 1 REGISTER FACT BOOK – FORMULA 3 (1970-72, 1973-77, 1978-81),** St. Leonard's Press, 1988.
STANLEY, LOUIS. **GRAND PRIX, THE 1964 CHAMPIONSHIP.** Doubleday, 1965
STARKEY, JOHN; WELLS, KEN. **LOLA THE ILLUSTRATED HISTORY 1957 TO 1977**. Veloce Publishing, 1998
THE NEW ILLUSTRATED ENCYCLOPEDIA OF AUTOMOBILES, The Wellfeet Press, 1992
TOMMASI, TOMMASO. **FROM INDIANAPOLIS TO LE MANS**. Derbibooks, 1974
VACARELLA, NINO. **IL PRESIDE VOLANTE: LA MIA STORIA AUTOMOBILISTICA**. Vacarella, 2013.
YOUNG, EOIN S. **FORZA AMON**, Haynes Publishing, 2003.

Several websites:
Autoracingcommentary.blogspot.com
Brazilexporters.com/blog
Carlosdepaula.blogspot.com
Statsf1.com
Wikipedia.org

MAGAZINES
The following magazines and newspapers have been consulted and quoted at one point or another. Clippings of a large number of unidentified newspapers and magazines were also consulted.
Auto Esporte
Auto Motor Und Sport
Auto Racing Digest
Grand Prix
Auto Week
Auto Zeitung
Automobilsport

Autosport
Autosprint
Car and Driver
Formula
Motor Sport
Motor Trend
Motoring News
On Track
Parabrisas Corsa
Quatro Rodas
Road & Track
Sport Auto

Made in the USA
Las Vegas, NV
10 December 2023

82540313R00177